Advances in Anatomy Embryology and Cell Biology

Vol. 131

Springer
Berlin
Heidelberg
New York
Barcelona
Budapest
Hong Kong
London
Milan
Paris
Tokyo

J. Matthias Starck

Comparative Anatomy of the External and Middle Ear of Palaeognathous Birds

With 57 Figures and 4 Tables

Springer

Dr. J. Matthias Starck
Eberhard-Karls-Universität Tübingen
Zoologisches Institut
Auf der Morgenstelle 28
72076 Tübingen, Germany

ISBN 3-540-58991-0 Springer-Verlag Berlin Heidelberg New York

Library of Congress Cataloging-in-Publication Data. Starck, J. Matthias, 1958– Comparative anatomy of the external and middle ear of palaeognathous birds/J. Matthias Starck. p. cm.—(Advances in anatomy, embryology, and cell biology; v. 131) Includes bibliographical references (p.) and index. ISBN 3-540-58991-0 1. Ratites—Anatomy. 2. Ear—Anatomy. I. Title. II. Series. QL801.E67 v. 131 [QL697] 574.4 s—dc20 [598.5′0448] 95-8260

Typesetting: Best-set Typesetter Ltd., Hong Kong

SPIN: 10474714 27/3136/SPS - 5 4 3 2 1 0 - Printed on acid-free paper

Contents

1 Introduction

1.1 Objectives of this Study

The vertebrate middle ear has attracted the interest of morphologists for more than a century. Its difficult structure, its complicated evolutionary derivation, and its integration of branchial, cranial, and otic materials into a single functional unit have made it a key organ for the understanding of vertebrate structures and their evolutionary history. Gaupp's (1898, 1913) and Reichert's (1837) comparative morphological studies of the vertebrate middle ear represented milestones for anatomy in the general recognition and acceptance of Darwin's theory of evolution. These fundamental studies notwithstanding, today's knowledge of avian middle ear structures is still characterized by descriptive studies focusing on character sampling to elucidate high-level phylogeny. Phylogenetic studies have considered either structural aspects of the bony stapes exclusively (Feduccia 1974, 1975a,b, 1976, 1977, 1978), or focused on the anatomy of the middle ear cavity, neglecting the sound transmission apparatus (Saiff 1974, 1976, 1978a,b, 1981, 1982, 1983, 1988). Other studies have investigated late-stage embryos and concentrated on the development of the skull, considering middle ear structures only as a side aspect. However, there are considerable structural differences between the middle ears of late-stage embryos, hatchlings, and adults of the same species. Although vertebrate morphology requires a meticulous knowledge of comparative middle ear data and calls upon an elaborate system of homologies, it turns out that knowledge of middle ear structural details is widely dispersed among different species and different developmental stages, making a comparison even more difficult.

In contrast to morphological studies, physiological investigations have concentrated on a few easily accessible species (chicken, barn owl, and budgerigar). They concern input-output relationships rather than structural functioning. Although a considerable amount of data has been compiled about structural and functional details of the middle ear, we still lack a thorough understanding of how its structural properties, its material components, and its function interact as an organ system specialized in sound transmission.

The objectives of this study are (1) to present detailed and new structural data regarding paleognathous birds and button quails (Turnicidae); (2) to present an interpretation of structures, their quantitative aspects, and their material components, which may help develop functional hypotheses; (3) to investigate the evolutionary history of these species; and (4) to contribute to a phylogenetic analysis by comparing these six paleognathous species with a neognathous species.

1.2 Literature Review

The literature concerning the avian middle ear can be classified according to the following approaches: (1) comparative morphological analyses, (2) phylogenetic analyses, and (3) functional analyses. The homologies of the avian middle ear and the hyoid arch of ancestral vertebrates have been established by Gaupp (1898, 1913), Schestakowa (1934), and more recently, on embryological evidence, by Smit and Frank (1979). Krause (1901) presented a comprehensive survey on the bony columella of 70 bird species. Freye-Zumpfe (1952, 1953) and Freye (1952, 1953) compiled basic comparative data on the avian middle ear. Outstanding examples are the comparative anatomical studies by Norberg (1977, 1978) on Tengmalm's owl, and the integration of anatomical and physiological data on the sound transmission apparatus in the budgerigar by Saunders (1985). More recent reviews on avian hearing mechanisms by Kühne and Lewis (1985) and Manley (1990) have concentrated on inner ear function and its central neuronal representation and do not provide new data or observations on middle ear structures.

Most comparative anatomical data have to be extracted from widely scattered references concerning the embryonic development of the skull. Since these studies do not focus primarily on middle ear structures, the quality of information differs considerably. It is beyond the scope of this paper to review all craniological publications. However, a comprehensive list of craniological references that provide comparative data on middle ear structures is presented in Table 1, which also identifies the species studied in each work. The comparative discussions in each section of this paper will refer to details from these studies.

Table 1. Craniological studies providing middle ear data

Struthionidae	*Struthio camelus*	Parker (1866a)
		Brock (1937)
		Lang (1956)
		Frank (1954)
		Webb (1957)
		Frank and Smit (1976)
		Saiff (1978a, 1981)
Rheidae	*Rhea americana*	Parker (1866)
		Müller (1961, 1963)
		Saiff (1983)
Dromaiidae	*Dromaius novae-hollandiae*	Parker (1866a)
		Lutz (1942)
Casuariidae	*Casuarius bennetti*	Parker (1866a)
		Lutz (1942)
		Kesteven (1942)
Apterygidae	*Apteryx australis*	Parker (1866a)
		Parker (1888, 1891)

Table 1. *Continued*

	Dinornis	Starck (1955)
	Apteryx australis, A. oweni	Saiff (1982)
Tinamidae	Several species	Parker (1866b)
	Several species	Saiff (1988)
Spheniscidae	Several species	Crompton (1953)
	Spheniscus demersus	Saiff (1976)
		Smit and Frank (1979)
		Anisimov and Iljitschew[a] (1980a)
Podicipedidae	*Podiceps cristatus*	Toerien (1971)
Procellariidae		Saiff (1974)
Phalacrocoracidae	*Phalacrocorax carbo*	Slaby (1951)
Pelecaniformes	Several species	Saiff (1979)
Ciconiiformes	Several species	Slaby (1952)
		Saiff (1978b)
Anatidae	Several species	de Beer and Barrington (1934)
		Schestakowa (1934)
		Swart (1946)
Falconidae		Suschkin (1899)
Phasianidae	*Gallus domesticus*	Parker (1869, 1890)
		Tonkoff (1900)
		Smith (1905)
		Sonies (1907)
		Pohlmann (1921)
		Jollie (1957)
		Saunders (1985)
		Vorster (1989)
		Cohen et al. (1992a,b)
Columbidae	*Columba livia*	Saunders (1985)
Rallidae	*Fulica atra*	Macke (1969)
Alcidae		Kartaschew and Iljitschew (1964)
Laridae	*Larus ridibundus*	Weber (1990)
Psittacidae	*Melopsittacus undulatus*	Lang (1954)
		Saunders (1985)
		de Kock (1987)
Strigidae	*Strix aluco*	May (1961)
	Aegolius funereus	Norberg (1977, 1978)
		Anisimov and Iljitschew (1980b)
Caprimulgidae	*Nyctisyrigmus pectoralis*	Fourie (1955)
Coliidae		Schoonees (1963)
	Colius indicus	Goldschmid (1972a–c)
Picidae	Several species	Parker (1875)
Sturnidae	*Sturnus vulgaris*	de Kock (1955)
Muscicapidae	*Turdus merula*	Slaby (1951)
Ploceidae	*Euplectes spec.*	Engelbrecht (1958)
	Passer spec.	Saayman (1963)
	Serinus canaris	Saunders (1985)
	Molothrus ater	Saunders (1985)
Corvidae		Anisimov and Iljitschew (1980)

[a] Several different transliterations may be found for Iljitschew (Iljitschow, Iljischow, Ilyichev).

A different spectrum of species was investigated in functional anatomical and physiological studies. Schwartzkopff (1955, 1957) described the amplification function by measuring the area ratio of the tympanic membrane and the clipeolus in 47 bird species. The auditory recesses of birds have received considerable attention in auditory research as interaural pathways since they might function as sound pressure gradient receptors during sound localization (Schwartzkopff 1950, 1952a,b; Payne 1971; Knudsen and Konishi 1979; Knudsen 1980; Hill et al. 1980; Coles et al. 1980; Rosowski and Saunders 1980; Lewald 1990). Several authors have attempted to describe the transfer function of the middle ear: Norberg (1978), Gaudin (1968), Tonndorf and Khanna (1970), Saunders and Johnstone (1972), Manley (1972), Gummer et al. (1989a,b), and Starck (1992) used various mechanical models. However, none of these models closely correspond to structural and physiological data, as shown by Saunders (1985). Because we know almost nothing about tissue properties and transfer parameters of middle ear structures, a considerable gap exists between the structural and functional data.

1.3 Theoretical Concepts: Questions and "Answers"

This paper deals with the comparative structural and phylogenetic analysis of the middle ear of paleognathous birds. The continuous struggle of morphologists with comparative methods, functional interpretation, and evolutionary history makes it necessary to briefly review some theoretical concepts that have influenced the preparation of this study.

Any investigation of structures faces three major difficulties: (1) Structural data are not self-evident; they are not a priori systematic, phylogenetic, functional, or evolutionary. To extract results from anatomical data, they must be analyzed by confrontation with theoretical concepts or physical laws (see following), and must be interpreted according to some theory. For this reason, and because the results are not independent of the researcher's experience, personal history, or social background, the method of analysis must be made explicit. (2) Descriptive anatomy, especially of rare specimens, is scarcely reproducible. In most studies, including the present one, only one or a few specimens of a species have been studied, depending on the availability of material and the conservation of endangered species. A thorough presentation of the structures and of the techniques applied is therefore necessary to understand the data and their origin. Relating structural data to physical properties of tissues, ecological observations on species, and the concept of evolutionary theory leads to functional, ecological, or evolutionary interpretations in morphology. Finally, (3) most comparative anatomical studies follow the inductive method, gathering data without prior formulation of hypotheses, which in turn makes it difficult to test and falsify their results.

One must differentiate between techniques, methodology, and interpretation. Techniques for the investigation of structures include dissection, histology, X-ray computed tomography, reconstruction, and description; they produce structural data and artifacts. Methodology refers to the analysis of the structural data, methods of comparison, statistics, or design of physiological experiments;

it depicts relationships and similarities between data and may also produce error. Finally, interpretation and evaluation consists of the confrontation of results with theoretical concepts.

1.3.1 Description and Representation

Homberger (1988) asked, "How can we make sure that what we see and describe is correct, i.e., represents reality?" There is no answer yet! According to the widely accepted criteria established by Popper (1969), scientific progress is based on hypotheses, their testing and experimental falsification, and the reproducibility of results (Popper 1969). In morphology, description is hardly reproducible, and redescription almost certainly results in new or differing data, based on different questions, different technologies, or different interests on the part of the investigators. There are no guidelines for deciding whether the first, the second, or both descriptions were right or wrong. Thus, descriptive anatomy risks not meeting an important criterion of science.

Homberger (1988) suggests two solutions for this dilemma. (1) In human and veterinary sciences, erroneous descriptions automatically have dreadful consequences. However, descriptions in vertebrate morphology have no such built-in corrective. In a strict sense they escape testing and reproducibility and cannot be falsified. (2) Homberger (1988) conceived a theoretical system for the testing of structural data: all structures are presumed to have functions which are based on their physical properties. Therefore, she concludes, a structure that can be described verified by relating it to its functional aspects. This concept entails many problems, since it is almost impossible to verify an analysis, rather than refute it. Furthermore, testing for the physical laws on which functional hypotheses are based must result in circular logic. Thus, the reliability of anatomical descriptions can hardly be tested by confrontation with a functional hypothesis, especially when the formulation of functional hypotheses is the goal of the study. Functional hypotheses can only be tested by experiments.

Here, I have chosen a more pragmatic approach. To improve the representation of structures, the birds studied have been investigated with a variety of microscopic, macroscopic anatomical, and noninvasive methods (see chapter 2). From all species, at least one hatchling and one adult specimen have been investigated in minute detail with various techniques. The use of various techniques and variously aged specimens should help avert methodical and individual artefacts.

1.3.2 Comparative Approach, Homology, Systematics, and Phylogeny

The comparative method, although implicitly the basis for all historical sciences, has only recently gained a theoretical foundation. Discussions of the comparative method by Dullemeijer (1974), Bock (1989), and Harvey and Pagel (1991) are sometimes controversial but cover most aspects of the comparative method, from comparison of discrete data to continuous variables, and from historical to nonhistorical comparisons. Here, we are concerned with discrete structural data which, however, are not independent of each other because of phylogenetic relationships between species.

Concepts of homology are diverse and controversial (Remane 1952, 1989; Osche 1973; Bock 1989; Ax 1989). Here, the concept introduced by Bock (1989) is used, which describes homologous structures in two or more organisms as "those that stem phylogenetically from the same structure (or condition) in the immediate common ancestor of these organisms." Consequently, phylogenetic analysis recognizes or describes homologies and not the other way around. Furthermore, homology does not depend on similarity of structures. This diversity and controversy among homology concepts must be kept in mind in the comparative discussion.

As mentioned previously, phylogenetic analysis is essential for this kind of comparative study since it clarifies the degree of relationship and independence in the data. Here, we can refer to the well-established monophyly of birds (Mayr and Amadon 1951; Cracraft 1981, 1986; Bock 1969) and the monophyly of paleognathous birds (Meise 1963; Bock 1963; Cracraft 1974; Bock and Bühler 1990; Houde 1988; Sibley and Ahlquist 1990; Sibley and Monroe 1990; Sibley et al. 1988).

Three equally valid, competitive theoretical concepts have been established for the phylogenetic analysis of characters: (1) *Evolutionary systematics* (Mayr 1969, 1974, 1990; Bock 1977) considers the similarities and differences between structures for an evolutionary classification. Evolutionary systematicists do not avoid classifications into paraphyletic groups, e.g., "reptiles," since all reptiles are more similar to each other than crocodiles are to birds. (2) *Cladistics* ("Konsequent phylogenetische Systematik"; Hennig 1950) establishes a system based exclusively on apomorphy of characters within a group. Phylogeny is based on synapomorphy of characters but not on similarities. Every structure-producing hierarchy can be used as a character in phylogenetic systematics. (3) *Numerical systematics* (Sokal and Sneath 1963) considers the percentile similarities between taxa. It does not consider ancient or derived characters. All three concepts provide useful and realistic descriptions of relationships between organisms. All use structures as guidelines for hierarchical classification and attempt to represent phylogeny.

1.3.3 Functional Approach

Structural data are not self-explanatory in a functional sense. As suggested by Bock (1991), structures can be confronted with physical properties and laws to derive hypotheses about their function (nomological-deductive approach; Bock 1991). This procedure leads to hypotheses about the form-function complex (Bock and von Wahlert 1965; α-function; Bühler 1987), which may be tested in further studies. Homberger (1988) suggested a similar procedure to derive functional considerations. Her approach, which expands this concept, derives a *structural model* from the anatomical description: "A structural model is a static representation of the essential structural characteristics of the functionally relevant elements of a complex mechanical system . . ." A *functional model* can in turn be derived from this structural model using the known physical principles; "a functional model is essentially the dynamic version of a structural model. It can be represented by a lever system, with acting and reacting forces of muscles and ligaments as in a structural model." The development of a functional model resembles the formulation of a hypothesis about the form-function complex (Bock and von Wahlert 1965) or Bock's nomological-deductive explanation (1990). In all cases, hypotheses about form-function interactions are tested experimentally or by independent functional observations. Natural variations in the structure under investigation, embryological malformations, and dysfunctions may also give some insights into the function of structures and are therefore of considerable interest for the formulation of a functional hypothesis; however, they are not as informative as experiments. Unfortunately, it is impossible to follow this approach while investigating middle ear structures since the acoustic and physical properties of bone, cartilage, and connective tissue are as yet poorly understood.

1.3.4 Ecomorphology

The physical functioning of a structure and its specific interaction with the species-specific environment are subsumed under the concept of ecomorphology (Bock and von Wahlert 1965; Leisler and Winkler 1985, 1991; Bock 1990). The interaction between the form–function complex and its specific environment are described as its *biological role* (*β*-function according to Bühler 1987). The detailed knowledge of a form–function complex is a necessary prerequisite to the study of its *β*-function. Furthermore, a detailed knowledge of the specific environment of the organism under investigation is necessary. Since the biological role of a structure is often difficult to prove experimentally, multivariate statistics can provide evidence for the most probable solution.

The materials and techniques used in this study do not provide insights into the biological role as described above. It would be rather trivial to describe the function of the middle ear as that of gaining acoustic information from the environment. However, to account for their acoustic environments, brief species descriptions will be given in section 3.1.

1.4 Basic Functional Concepts of the Avian Middle Ear: The Transfer Function

In a generally accepted view (Gaudin 1968; Norberg 1978; Gummer et al. 1989a,b; Manley 1990; Starck 1992), the avian middle ear functions as a sound-conducting apparatus, transducing airborne sound into movements of inner ear fluid. Matching of the high acoustic impedance between air and fluid determines the form-function complex of the middle ear. Low-pressure, high-velocity sound in the external auditory meatus must be transformed into high-pressure, low-velocity vibrations in the inner ear. An amplification function of ca. 63 is necessary for the impedance between external and inner ear media to match (Møller 1965).

All structures, including such "nonstructures" as air spaces and connective tissue layers, may contribute to the amplification of the transfer function. However, they also contribute to the impedance of the middle ear. The structural aspects of amplification and impedance of the middle ear will be termed "morphological amplification function" and "morphological impedance," respectively.

Four parameters are thought to contribute to the morphological amplification function: (1) the ratio of the tympanic membrane to the clipeolus; (2) the lever system constituted by the columella and extracolumella; (3) the curvature of the tympanic membrane, contributing to a lever function known as the curved membrane effect; and (4) the air-filled spaces of the middle ear and the adjacent pneumatic spaces of the skull, which may function as resonators and contribute to frequency-dependent amplification.

The morphological impedance depends on: (1) mass reactance of all structures involved in the transfer function; (2) stiffness reactance, referring to the compliance of the system; and (3) mechanical friction.

1.4.1 Area Ratio

The area ratio is an easily measured parameter of the middle ear that can be obtained by planimetry of the tympanic membrane and clipeolus areas. A series of measurements on neognathous birds (Schwartzkopff 1955, 1957) gives area ratios between 15 and 40. However, physiological investigations present evidence that the actually measured area of the tympanic membrane is not the effective area. It has been shown in mammals (Tonndorf and Khanna 1970) and reptiles (Manley 1990) that the oscillation pattern of the tympanic membrane changes with the frequency of the sound received. At high frequencies (>3 kHz), the tympanic membrane as well as the extracolumella reportedly lose impact on the transfer function (Manley 1972, 1990).

It is highly probable that similar oscillation patterns could be described in birds. The extracolumella inserts peripherally into the tympanic membrane, indicating a functional area different from the actual tympanic area. Furthermore, the perilymphatic surface of the clipeolus is not always a plane. As described by Krause (1901), the perilymphatic surface of the clipeolus of many avian species (e.g., barn owl) bulges into the perilymphatic recess and changes the area ratio and the transfer function. Thus, any morphological measurement represents only an estimate of the effective area ratio of tympanic membrane and clipeolus. The effective area of the tympanic membrane of birds has been estimated to be 70% of the anatomical area (von Békésy 1960). However, since these estimates have remained without further confirmation, this paper will refer to the original anatomical measurements.

1.4.2 Lever Systems

Several lever systems have been suggested for the middle ear of birds (Gaudin 1968; Norberg 1978; Saunders 1985; Gummer et al. 1989a; Cohen et al.

1992a,b; Starck 1992). The avian middle ear is considered a single-ossicle lever system, with the cartilaginous extracolumella, its three processes, and the bony stapes as functional components. Controversies about the suggested lever systems concern the position of the fulcrum, the definition of force lever and resistance lever arms, and the hypothetical movements of the clipeolus. Some papers based on anatomical evidence (Gaudin 1968; Cohen et al. 1992a) suggest a movement of the clipeolus around the stiffer hinge point of the annular ligament at the posterior end of the oval window. Others (Norberg 1978; Starck 1992) suggest piston-like movements of the clipeolus. Direct physiological measurements (Saunders 1985; Gummer et al. 1989a,b) support the latter view. As already discussed in the previous paragraph, a considerable difference exists between the lever ratios calculated on the basis of anatomical measurements and those found by physiological input-output measurements. Gummer et al. (1989a,b) calculated a geometric lever ratio of 1.6 for the stapedial complex in the pigeon. However, their physiological measurements revealed a mechanical lever ratio of 2.7.

1.4.3 Curved Membrane and Resonance Effects

The tympanic membrane itself acts as a lever system, as described by von Helmholtz (1868). This lever system depends on the curvature of the tympanic membrane, its area, and the extent to which the tympanic process displaces the tympanic membrane. Its effect, however, is difficult to assess. Tonndorf and Khanna (1970) measured the curved membrane effect on the tympanic membrane of the cat and found it to contribute a factor of two to the overall amplification function. The effect of the curved membrane has not yet been investigated in birds.

Resonance of the large air-filled cavities in the skulls of most birds is thought to contribute to a frequency-dependent amplification of the sound transfer. However, it is difficult to calculate the resonance frequencies and their proportion of the overall amplification function. A simplified model considers the avian middle ear as a Helmholtz resonator. Although this is only a rough approximation, it at least provides an estimate of the resonance properties of the middle ear and its adjacent pneumatic spaces.

1.4.4 Impedance of the Middle Ear

Several parameters contribute to the morphological impedance of the middle ear. Their effects are frequency-dependent; thus, it is impossible to determine their contribution to the transfer function by structural analysis alone (Kühne and Lewis 1985). However, morphometry can provide some data on structure-dependent impedance of the middle ear: (1) *Mass reactance* refers to the masses of all structures involved in sound transfer and their influence on the transfer function. It changes proportionally with frequencies. (2) *Stiffness reactance* refers to systemic compliance. It depends not only on the compliance and material components of the structures themselves but also on the volume of the air cushion enclosed by the middle ear cavity and the adjacent pneumatic spaces. Stiffness reactance acts reciprocally to frequencies. (3) *Mechanical friction*, which is a constant proportion of the total morphological impedance, is independent of frequency.

1.5 Evolutionary History of Middle Ear Structures

1.5.1 Homologies of the Avian Middle Ear and the Hyoid Arch

A long discussion in comparative vertebrate morphology concerns the homologies of the auditory ossicles. I will briefly review these positions as far as they concern the avian middle ear. As mentioned previously, the term homology is consistently based on the recognition of immediate common ancestry. If homology is used with any other meaning in this paper, it will appear in quotation marks. The homology discussion is reviewed here only to provide necessary background information and to depict that portion of the avian middle ear that is designed by historical constraints rather than by functional optimization.

Since Gaupp's (1913) classical work, a broad consensus in comparative anatomy relates the sound-conducting apparatus of the amniote middle ear to the hyoid arch of the crossopterygian branchial skeleton. This general "homology" is based on embryological evidence indicating that the cartilaginous precursors of the sound-conducting apparatus are derived from mesenchyme of the second embryonic branchial arch (hyoid arch). Specifically, the regional "homology" of the hyoid arch and the sound-conducting apparatus is based on (1) the argument of the topographic relationships between the hyoid arch in early-stage embryos and the middle ear of the adult (e.g., the nervus facialis, chorda tympani, arteria carotis interna, and musculus stapedialis, and the topographic relationship to the otic capsule), and (2) the argument of continuity in development from the embryonic hyoid arch to the sound-conducting apparatus in adult vertebrates.

Despite this general consensus, there is considerable disagreement concerning the "homologies" of different parts of the columella and how these relate to the crossopterygian hyoid arch. Furthermore, confusion arises from the question as to whether the tympanic membrane is homologous in all vertebrates or has evolved several times independently.

1.5.1.1 Stapes

In earlier studies, the stapes has been considered a structure "homologous" to the pharyngohyal element of the hyoid arch (Versluys 1904; Schestakowa 1934). Recent comparative studies by Frank and Smit (1974, 1976) and Smit and Frank (1979) showed the composite nature of the stapes. Their embryological investigations of birds and crocodiles revealed two mesenchymal blastemas, which they described as suprapharyngohyal and infrapharyngohyal elements. The infrapharyngohyal blasteme is thought to contribute all material to the stapes. The suprapharyngohyal contributes material to the distal end of stapes and some material to the supracolumellar process of the extracolumella.

There has been considerable dispute over whether the stipes (shaft of the columella) and the clipeolus develop from the same origin. Earlier papers suggested that the clipeolus and the stipes derive independently, with the clipeolus as "otostapes" from the lateral wall of the otic capsule and the stipes as "hyostapes" from the hyoid arch (Hoffmann 1889; Suschkin 1899). Others, however, present evidence for a common origin of stipes and clipeolus from exclusively hyoid material (Sonies 1907; de Beer and Barrington 1934; Schestakowa 1934; Lutz 1942; Toerien 1971; Goldschmid 1972a–c; Frank 1954; Frank and Smit 1974, 1976; Smit and Frank 1979; Vorster 1989). Crompton (1953), Slaby (1951), and Engelbrecht (1958) report an independent chondrification of the clipeolus and the stipes. These authors again cannot exclude contribution of material from the otic capsule to the clipeolus.

1.5.1.2 Tympanic Process of the Extracolumella

The tympanic process of the extracolumella has been considered "homologous" to the ventral part of the hyomandibula. Schestakowa (1934) described it as a "symplecticum." This "homologization" is based on a mesenchyme connection between the tympanic process and the mandibula in the duck. However, the symplecticum is a ventral ossification in the hyomandibula of actinopterygian fish that connects to the braincase and supports the lower jaw. It is not found in any ancestral tetrapod. Thus, its direct "homologization" with the cartilaginous tympanic process of the avian columella is impossible, even with a traditional view on "homologies."

Frank and Smit (1974, 1976) and Smit and Frank (1979) found the tympanic process to be composite in nature. Its proximal part is described as belonging to the infrapharyngohyal blasteme. Contributions of material derived from the epihyal blasteme and the suprapharyngohyal blasteme are reported for the middle part of the tympanic process, whereas its distal part is thought to be a derivative of secondarily delaminated laterohyal material. It is important to note that in adult birds, the border between the bony stapes and cartilaginous tympanic process of the extracolumella is not equivalent to the border between infrapharyngeal and laterohyal material. Thus, with reference to embryonic orgin, the later intracolumellar articulation has an intrainfrapharyngeal position.

1.5.1.3 Processus supracolumellaris

There are considerable differences between the "homologizations" of the supracolumellar process. Schestakowa (1934) described the supracolumellar process as "homologon" to the hyomandibula. She refers to its position between the parotic process and quadrate and lateral to the lateral head vein, as well as to its relative position within the chain of blastemas. A later reinvestigation by Frank and Smit (1974, 1976) again suggests that the supracolumellar process is composite in nature. The medial prong of the supracolumellar process is derived from the suprapharyngeal element, whereas the lateral prong is related to laterohyal material.

1.5.1.4 Processus infracolumellaris

The infracolumellar process has been described as being derived from epihyal material, hyomandibula (Schestakowa 1934), interhyal (stylohyal) material, and ceratohyal material (Frank and Smit 1974, 1976; Smit and Frank 1979). It has no direct connection to the lower jaw, although the ventral tip of the infracolumellar process may come close to the internal process of the mandible. Arguments for a "homologization" are based on the topographic sequence of embryonic blastemas within a chain of branchial blastemas, and the lack of a connection between the infracolumellar process and the lower jaw. Logically, its "homologization" can be considered a consequence of the previous "homologization" of the supracolumellar process with a hyomandibula (Schestakowa 1934) or with suprapharyngeal and laterohyal elements (Frank and Smit 1974, 1976).

An examination of the embryonic structures in birds and in crossopterygian and actinopterygian fish shows nothing more than that the avian sound-conducting apparatus develops from the same embryonic region (hyoid arch region) as the jaw suspension apparatus in fish. Thus, assuming a common vertebrate ancestor, a regional homology can be recognized. On a more detailed level of comparison, the avian columella and the jaw suspension apparatus in fish have nothing in common. Table 2 summarizes the "homologies" discussed above (see Gadow 1891, p. 475, for a summary of the nineteenth century discussion of the homologies of the hyal arch). I have not attempted to add

Table 2. Synonyms and homologizations for the avian stapes and extracolumella

	Schestakowa (1934)	Crompton (1953)[a]	Frank and Smit (1974, 1976) Smit and Frank (1979)
Stapes	Pharyngohyal	Pharyngohyal-Epihyal	Infrapharyngohyal
Proc. tympanicus	Symplecticum (= ventral part of hyomandibula)	Epihyal (ventral part)	Infrapharyngohyal Laterohyal material
Proc. supracolumellaris	Dorsal part of hyomandibula	Epihyal	Suprapharyngohyal Laterohyal material
Proc. infracolumellaris	Epihyal (= Hyomandibula) Interhyal (Stylohyal) Ceratohyal	Ceratohyal Hypohyal	Interhyal Ceratohyal

Proc., processus.
[a] Crompton (1953) explicitly states that the terms "pharyngohyal," "epihyal," "ceratohyal," and "hypohyal" should not imply homologies to the same elements in the crossopterygian hyal arch.

any new hypotheses concerning homologies of the avian columella and extracolumella.

1.5.2 Phylogenetic Relationship of Palaeognathous and Basic Neognathous Birds

Comparative studies depend on the phylogenetic relationships of the taxa considered. A short review of current positions is provided to help evaluate the structural data presented here. It is of merely historical interest that some authors considered birds to be polyphyletic. Lowe (1926, 1928, 1944), Holmgren (1955), Glutz von Blotzheim (1958), and Verheyen (1960a,b) related birds to various stem groups of Saurischia. Their arguments are orthogenetic and refer to the structural aspects of flightlessness in "ratites," which were thought to represent an ancestral state. Today, a broad consensus considers birds to be monophyletic, characterized by an unquestionable set of apomorphies, and derived from theropods as their common ancestor (Bock 1969; Cracraft 1981, 1986, 1988; Currie 1985; Gauthier and Padian 1985; Houde 1988; Sibley et al. 1988; Cracraft and Mindel 1989; Sibley and Ahlquist 1990; Sibley and Monroe 1990; Witmer 1990; Chatterjee 1991; Elzanowski 1994).

Phylogeny within the Aves is more confused. Since the description of the palaeognathous palate by Huxley (1867) (dromeognathous, in his terminology), the high-level phylogeny of birds has been vigorously debated. However, the discussion refers to various levels and concepts of phylogeny. The monophyly of palaeognathous birds is widely accepted. On the basis of cranial structural findings and principles of evolutionary classification, Bock (1963, 1969) presents strong evidence for the monophyly of palaeognathous birds (Struthioniformes). His main argument refers to the complexity of the palaeognathous palate structure: "The features characterizing the palate in all groups of ratites are so numerous that a conclusion other than common ancestry in scarcely conceivable." Parkes and Clark (1966) support the monophyly of palaeognathous birds (Ratitae and tinamous) with evidence from the structure of the rhamphotheca. In a later paper, Bock and Bühler (1990) present additional anatomical evidence from the tongue apparatus for the monophyly of palaeognathous birds. However, they show that ratites (as flightless palaeognaths) are polyphyletic and that flightlessness evolved along several independent evolutionary pathways in palaeognathous birds. The classification of palaeognathous birds as suggested by Bock and Bühler (1990) is given in Table 3. It must be pointed out that Bock and Bühler (1990) draw no conclusions about the phylogenetic relationship of palaeognathous to neognathous birds. Recently, Elzanowski (1994) presented a phylogenetic analysis including mesozoic bird material and considered the palaeognathous birds a monophyletic sister taxon to recent neognathous birds. The palaeognathous palate, the monocondyle quadratocranial articulation, and the elongated zygomatic process are considered palaeognathous synapomorphies (see 4.2 for a discussion of these characters). Unfortunately, his study does not consider all palaeontological evidence (e.g., Protoavis, Chatterjee 1991).

Cracraft (1974, 1981, 1986, 1988) investigated the postcranial anatomy of palaeognathous birds and used a cladistic approach for the reconstruction of

Table 3. Classification of Struthioniformes according to Bock and Bühler (1990)

Order Struthioniformes (Palaeognathae)
Suborder Tinami
Superfamily Tinamoidea Gray, 1840 (1831)
Family Tinamidae Gray, 1840 (1831)
Family Rheidae Bonaparte, 1849
Superfamily Apterygoidea Gray, 1840
Family Dromaiidae Huxley, 1868
Family Casuariidae Kaup, 1847
Family Dinornithidae[a] Bonaparte, 1853
Suborder Struthioni
Family Struthionidae Vigors, 1825
Family Aepyornithidae[a] Bonaparte, 1853

[a] Extinct taxon

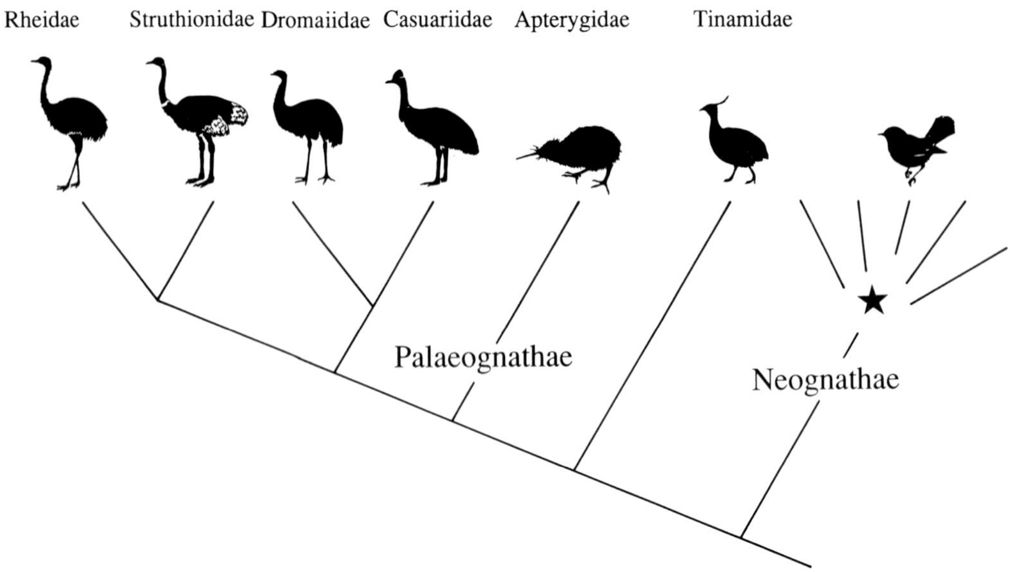

Fig. 1. Phylogeny of palaeognathous birds based on Cracraft (1974)

their phylogeny (Fig. 1). The monophyly of palaeognathous birds is based on the palaeognathous palate, a unique ramphothecal structure, and a large ilio-ischiadic fenestra. Using outgroup comparison and functional interpretation of the pelvic structure in flightless birds, Cracraft (1974, 1981) discussed the ilioischiadic window as a derived structure. He considered the tinamous to be a sister group of all other palaeognathous birds (Ratitae). Derived characters of palaeognaths are features of the tibiotarsus, specializations in the pelvis, and fusion of the skeletal elements of the shoulder. Cracraft and Mindel (1989) provide a thorough discussion of recent phylogenetic hypotheses about the early history of modern birds based on various techniques (morphology, ribosomal DNA, α-crystallin, transferin ID, and DNA hybridization). Only

DNA hybridization and ribosomal DNA data suggest that palaeognathous birds originate within the neognathous birds, whereas hypotheses based on other techniques recognize them as sister groups (see Bock and Bühler 1990, for a controversial solution). On the basis of a "global parsimony solution" they suggest that Palaeognathae are a sister group of all other birds.

Meise (1963) presented an independent cladistic approach using behavioral and reproductive biology characters of palaeognathous birds. He also classified the Tinamidae as a sister group of all other palaeognathous birds. However, this is based on the egg structure only.

Houde (1988) reported new palaeontological evidence of palaeognathous birds. He described the Lithornithidae as a family of palaeognathous birds showing many tinamou-like cranial characters but cathartid features of the postcranial skeleton. In his phylogenetic analysis, extant palaeognathous birds are relict groups of a large tertiary radiation. He considers the lithornithids and (classic) Ratitae as sister taxa of tinamous and neognathous birds, which again represent sister groups. The palaeognathous palate is considered a plesiomorphic (primitive, in his terminology) condition.

Molecular biology techniques have provided independent data for phylogenetic analysis. Sibley et al. (1988), Sibley and Ahlquist (1990), and Sibley and Monroe (1990) presented a phylogenetic analysis based on $\Delta T_{50}H$ values of DNA-DNA hybridization techniques and showed a monophyletic taxon of palaeognathous birds. Their analysis supports the hypothesis of tinamous being a sister group of the Ratitae (Struthioniformes, in their classification). Struthioniformes and Tinamiformes (sensu Sibley et al. 1990), together with Galliformes and Anseriformes, are considered a monophyletic taxon, Eoaves, which is a sister taxon to all other birds (Neoaves). However, the acceptance of their phylogeny and classification depends largely on the evaluation of DNA-DNA hybridization techniques as a valid tool for phylogenetic analyses.

The monophyly of palaeognathous birds seems to be well documented on the basis of these studies. However, phylogenetic analysis depends not only on the methodology applied but also to a considerable extent on the evaluation of characters. Thus, some studies result in a contradictory classification of palaeognathous birds, even using the same material. In his study of the bony palate of birds, McDowell (1948) recognizes more dissimilarities than similarities among the palate structures of palaeognathous birds. He strongly argues for the independent evolutionary origin of all groups of palaeognaths. In his view, the palaeognathous palate either developed independently as adaptation to feeding mechanisms or is neotenic. The "neoteny argument" has been developed further by De Beer (1956), who considered the palaeognathous palate "an arrested stage in the development of the neognathous palate." Further neotenic characters are seen in the feather structure as well as the sutures of the dermal skull bones. However, De Beer avoids concluding that the palaeognaths are monophyletic. Simonetta (1960), in his interpretation of palaeognathous skulls, suggests that palaeognaths evolved along several independent pathways to acquire a less kinetic skull. He considers the palaeognathous palate to be an adaptation to feeding mechanisms. McDowell's (1948) interpretation of neotenic characters was rejected by Simonetta (1960). More recently, the neoteny argument has been revived by Storer (1971), Olson

(1985), and Feduccia (1980). They suggest several scenarios according to which either all palaeognaths are neotenous descendants of neognathous ancestors, or some palaeognathous taxa stem from an archaic group, whereas others arose independently through neoteny of neognathous birds and became secondarily palaeognathous. These studies do not identify the potential stem groups for the derivation of neotenous palaeognaths. The neoteny discussion has been reviewed by Elzanowski (1989), who argued that the hypothesis of a neotenic origin of ratites is based on a series of misconceptions of embryonic characters. The reader is referred to his paper for a detailed discussion and evaluation of all arguments used in the neoteny discussion.

Here, the classification by Bock and Bühler (1990) as summarized in Table 3 will be used as a reference for the phylogenetic analysis of middle ear data (see 4.2).

The classification of button quails (Turnicidae) is an open question in avian systematics. Mayr and Amadon (1951), Bock and McEvey (1969), and Bock et al. (1975) consider them a family within the gruiform birds. In the recent molecular classification by Sibley and Monroe (1990), button quails (Turnicidae) are treated as an order incertae sedis. They are placed tentatively at the first branching node within the Neoaves, and are considered a sister group of all remaining neognathous birds. In earlier papers, phylogenetic relationships have been postulated to pigeons and plovers (Parker 1866b), to Galliformes or to Columbiformes (Lowe 1926). Cracraft (1981) again considers them as incertae sedis, and Wolters (1975–1982) places them in a separate order, Turniciformes, close to the Gruiformes.

2 Materials and Methods

2.1 Species List

Species of all paleognathous families have been investigated in this study and compared to a basal neognathous species.

Rhea americana [Linneaus 1758] Rheidae, Struthioniformes: (1) Adult unnumbered specimen. Department of Anatomy, Faculty of Medicine, University of Frankfurt. (2) Hatchling, unnumbered specimen, histological serial sections through the head, celloidin, 40 μm, azan stain. Department of Antomy, Faculty of Medicine, University of Frankfurt. Courtesy of D. Starck.

Eudromia elegans [Geoffroy Saint-Hilaire, 1832] Tinamidae, Struthioniformes: (1) Adult specimen, collection of J.M. Starck, STA/EU 6. (2) Hatchling STA/EU, three histological serial sections through the head, paraffin, 15 μm azan stain.

Tinamous tao [Temminck 1815] Tinamidae, Struthioniformes: Unnumbered specimen, histological serial sections through the head of a late-stage embryo, celloidin, 30–40 μm, azan stain. Department of Anatomy, Faculty of Medicine, University of Frankfurt. Courtesy of D. Starck.

Dromaius novae-hollandiae [Latham 1790], Dromaiidae, Struthioniformes: (1) Adult specimen, collection of the American Museum of Natural History No. 7849. (2) Histological serial sections, celloidin, 30–40 μm, azan stain, through the head of a hatchling, unnumbered specimen (origin: Zoo Köln, FRG). Department of Anatomy, Faculty of Medicine, University of Frankfurt. Courtesy of D. Starck.

Casuarius bennetti [Gould 1857], Casuariidae, Struthioniformes: Adult specimen, collection of the Zoological Institute, University of Tübingen, unnumbered specimen.

Casuarius casuarius [Linneaus 1758], Casuariidae, Struthioniformes: Histological serial sections through the head of a hatchling, celloidin, 30–40 μm, azan stain, Department of Anatomy, Faculty of Medicine, University of Frankfurt. Courtesy of D. Starck.

Apteryx australis [Shaw 1813], Apterygidae, Struthioniformes: (1) Adult unnumbered specimen, Department of Anatomy, Faculty of Medicine, University of Gießen. Courtesy of H.-R. Duncker. (2) Histological serial section through the head of a hatchling, celloidin, 40 μm, azan stain. Collection of the Zoological Institute, University of Tübingen, no. SAI/78110 (origin: Zoo Frankfurt/Main, FRG; October 1978, 4 days old).

Struthio camelus (Linneaus 1758), Struthionidae, Struthioniformes: (1) Adult unnumbered specimen, Department of Anatomy, Faculty of Medicine,

University of Frankfurt. (2) Histological serial sections through the head of a hatchling, celloidin, 30–40 μm, azan stain. Department of Anatomy, Faculty of Medicine, University of Frankfurt (Origin: Zoo Basel, Switzerland; September 1953). Courtesy of D. Starck.

Turnix suscitator [Gmelin 1789], Turnicidae, order incertae sedis: (1) Adult specimen, collection of J.M. Starck STA/TU 7. (2) Histological serial sections, paraffin, 15 μm, azan stain, through the head of a hatchling no. STA/TU 132-205.

Gallus gallus [Linneaus 1758], Phasianidae, Galliformes: Adult specimens of domestic fowl were preserved for electron microscopy.

2.2 Macroscopic Dissections

Macroscopic dissections were made in as detailed and careful a manner as possible. All steps of gross dissection were documented with photographs and drawings. Tympanic membranes were cut at their margins, where the epithelial cover turns into the cover of the external auditory meatus. The tympanic membrane and columella were removed together.

2.3 X-Ray Computed Tomography (CT)

The air spaces in the tympanic cavity and the adjacent pneumatic spaces in the skull cannot be investigated using macroscopic dissection techniques. Any dissection would destroy the "non-structures" (air-filled spaces). X-ray computed tomography (CT) provides a powerful non-invasive technique, creating serial images for reconstruction and morphometry. Here, 1-mm CT scans were generated on a Phillips Somatom (Siemens, Erlangen, Germany) at the Department of Neuroradiology, Faculty of Medicine, University of Tübingen.[1]

2.4 Morphometry

Measurements during macroscopic dissections were taken with a slide caliper to the 1/10 mm. Tympanic membrane and clipeolus areas were measured on camera lucida drawings. Linear measurements of the columella and extracolumella were taken with a calibrated lens. Volumetry of the tympanic spaces and pneumatization of the skull were computed from the digitized CT scans. As shown previously (Zilles et al. 1982; Starck and Kriete 1989), measurements of 25 equidistant sections result in data with a systematic error below 1%.

[1] I gratefully acknowledge the help and support of Dr. Ozdoba during the computed tomography procedures.

2.5 Microscopic Investigations

2.5.1 Serial Sections

I investigated serial sections through the heads of hatchlings of all species. The heads of the larger species (ostrich, American rhea, emu, cassowary, and kiwi) were embedded in celloidin and sectioned in series of 30–40-μm-thick slices. The heads of the button quail hatchling and the crested tinamou hatchling were embedded in paraffin and sectioned to 15-μm slices. All series were stained with azan stain (Romeis 1989).

2.5.2 Transmission Electron Microscopy

Tissue samples of the tympanic membrane, Platner's ligament, and the ascending ligament were fixed in 2.5% glutardialdehyde in 0.1 mol Soerensen buffer (pH 7.4) for 2 days, washed in Soerensen buffer, immersed in osmium tetroxide 1% for 24 h, dehydrated, and embedded in epon. Ultrathin sections were studied on a transmission electron microscope (Siemens Elmiskop 102, Siemens, Berlin, Germany).

2.5.3 Three-Dimensional Reconstructions

The middle ear region and basicranium of hatchlings of all species was modeled from serial sections by styrofoam reconstruction. Cartilaginous and bony structures, nerves, arteries, and veins were drawn directly from serial sections on 2-mm-thick styrofoam sheets using a drawing tube (camera lucida). For reconstructions, a magnification between 25 and 50 was chosen, depending on the size of the original cross sections through the head. After drawing, the structures were cut and the styrofoam plates mounted. The complete models were photographed and drawn with the airbrush technique to present an optimal three-dimensional effect.[2]

2.5.4 Computer-Aided Three-Dimensional Reconstructions

Serial CT scans through the heads of the adult specimens were used for reconstruction of the air-filled spaces in the head. Magnified paper prints of the serial CT scans, the outline of the skull, and the air-filled spaces were manually digitized on a digitizing tablet. Depending on the size of the head, series of 25–73 CT scans were digitized. Three-dimensional reconstructions were generated on a PC 80386 using the Gred program (Kontron, Eching, Germany). This program also provides options for volumetry.

---

[2] All airbrush paintings were done by Margret Roser, Zoological Institute, University of Tübingen.

2.6 Nomenclature, Synonyms, and Abbreviations

Today, there is no consistent and generally accepted nomenclature of middle ear structures available for birds. The *Nomina Anatomica Avium* (NAA; Baumel et al. 1979, 1993) does not cover the sensory organs appropriately. Therefore, I am including a comprehensive list of terminology comprising all important synonyms to faciliate comparison with other studies on avian middle ear structures. This list is in alphabetical order according to the abbreviations used in the figures.

3 Results

3.1 Species Biology

Studying middle ear structures means investigating a sound transmission apparatus that receives acoustic information from the environment, conspecifics, other animals, or other acoustic sources, and transforms them into pressure waves of endocochlear fluid. Knowledge of the hearing range of each species is a necessary prerequisite for conclusions on functional morphology. To draw conclusions on the biological role of the middle ear structures, information about the acoustic behavior and the bioacoustic environment of the species should be taken into consideration. Thus, although the main goal of this paper is the descriptive morphology of middle ear structures, a short description of the habitat of each species will be given and the available information on acoustic behavior and the bioacoustics of their respective environments will be reviewed. For none of the investigated species is there a hearing threshold curve available.

3.1.1 American Rhea

The American rhea (*Rhea americana*; Fig. 2A) is the largest of the New World palaeognathous birds, weighing 20–25 kg and reaching a height of 1.4 m. It is a large, flightless bird inhabiting grassy plains and open brush in eastern and central Brazil, Bolivia, Paraguay, Uruguay, and southwestern Argentina, from low altitudes to up to 2000 m. It is cursorial, runs swiftly, and is said to swim well. The call of adult rheas is described as a deep, booming "bu-up" or "nau-du" (Folch 1992).

3.1.2 Crested Tinamou

The crested tinamou (*Eudromia elegans*; Fig. 2B), weighing about 600 g, belongs to the larger tinamou species [in the tinamou family, body mass ranges from ca. 80 g (*Taoniscus nanus*) to 1100 g (*Tinamous major*)]. The crested tinamou inhabits arid and semiarid regions, chiefly the dry savannahs of South America (Argentina and Chile). Occasionally, it is also found in open woodlands or dry Andean steppes. It is mostly confined to lowlands and Andean foothills but can also be found at elevations of up to 3500 m above sea level. Its acoustic behavior has not been described. From my experience with captive

A
B
C
D
E

Fig. 2. **A** American rhea (*Rhea americana*). **B** Crested tinamou (*Eudromia elegans*). **C** Emu (*Dromaius novae-hollandiae*). **D** Dwarf cassowary (*Casuarius bennetti*). **E** Kiwi (*Apteryx spec.*). **F** Ostrich (*Struthio camelus*). **G** Barred button quail (*Turnix suscitator*). Drawn from photographs of live birds (**A,B,F**) or preserved specimens (**C,D,E**); **G** photograph (J.M. Starck)

F

G

crested tinamous, at least two different sounds can be described. The first is a soft, booming note that lasts ca. 2 s and is uttered at regular intervals. Booming occurs in social and aggression-free situations and continues for several minutes. The other note is a high-pitched, noisy call that can be heard at some distance (100–200 m), even within buildings. These calls might be of some importance in courtship behavior.

For comparison, Skutch (1983) and Wetmore (1981) described the calls of the great tinamou (*Tinamous major*) and the little tinamou (*Crypturellus soui*). In contrast to the crested tinamou, both of these species inhabit the humid tropical South American lowland forest up to 1500 m and 1700 m. The great tinamou is described as having strong and full calls consisting of one or more phrases, each composed of a short note followed by a prolonged, ascending note (Skutch 1983). Great tinamous call throughout the day and night but especially in the early morning and late afternoon. The call of the little tinamou is described as a melodious, whistled song. It is repeated several times and varies considerably in volume, especially at the end, when it gains in strength and rapidity with each repetition (Skutch 1983; Wetmore 1981). Little tinamous answer each other. They call throughout the day and night.

3.1.3 Emu

The emu (*Dromaius novae-hollandiae*; Fig. 2C) is the second-largest palaeognathous bird. It reaches a height of 1.5–1.9 m and has a mean body mass of 30–35 kg. It is endemic to Australia, where it is widespread in open country, woodlands, the savannah, and the semidesert. It intrudes into township areas where it is sometimes treated as a pest, grazing on cultivated land. It is seen in pairs or small groups. The emu generally feeds on green vegetable matter and fruits, but also on insects and other animal prey. The voice of the emu is described as harsh and guttural in the male, and as booming in the female. Acoustic display has been observed during courtship and aggressive encounters. It can be heard up to 2 km away. Females utter a typical "e-moo, e-moo" during breeding season (Folch 1992). A trachea cleft leading into an expandable pouch, probably representing a resonance cavity, has been described in emu females (Murie 1867).

3.1.4 Cassowary

Cassowaries (Fig. 2D) are confined to the tropical rain forest of New Guinea and northern Queensland, Australia. The dwarf cassowary (*Casuarius bennetti*) is the smallest of the "large flightless" birds, reaching ca. 1.0–1.1 m in height and a body mass of ca. 17.6 kg. It can easily be distinguished from the other species by its casque, which is distinctly smaller, triangular, and flattened in the back. The dwarf cassowary has no wattles on its neck. It is found in the dense rain forests of the lowlands and mountains of New Guinea and New Britain from sea level up to 3300 m, where it feeds on fruits. Its call has been described as low, booming grunts, and also as a loud mwaaaa! (Beehler et al. 1986). Booming often occurs with a special posture termed "booming display" (Coates 1985), where the head is held down with the tip of the bill pointing backwards. During "booming display" the neck and head are inflated (see following).

The southern cassowary (*Casuarius casuarius*) inhabits New Guinea and a small region of tropical rain forest in Northern Queensland, Australia. It is taller than the dwarf cassowary, reaching 1.3–1.5 m (females 58 kg; males 29–34 kg). Feeding and habits are similar to those of the other cassowary species. Its call is described as a harsh "heugh-heugh" and a deep, booming sound (Beehler et al. 1986).

3.1.5 Kiwi

The three species, the brown kiwi (*Apteryx australis*), the little spotted kiwi (*A. owenii*), and the great spotted kiwi (*A. haastii*) are of relatively small body size (males 1500 g, females 2500 g; see Fig. 2E). Kiwis are flightless birds with extremely reduced wings. The kiwi is well known for its adaptation to nocturnal and crepuscular life. It has extremely small eyes and is said to have well-developed olfaction. Kiwis are endemic to the northern and southern islands of New Zealand. They live in humid forests whose soils consist of organic wastes, where they forage for insects and other invertebrates. The call of the kiwi is described as a short and sharp, shrill whistle or scream (Stresemann 1934; van Tyne and Berger 1976).

3.1.6 Ostrich

The ostrich (*Struthio camelus*; Fig. 2F) is the largest of all living birds. Males reach 2.1–2.7 m in height, females 1.75–1.9 m. Their weight can reach 130 kg. The ostrich inhabits the desert and savannahs ranging from south of the Sahara to the East African savannah as far as Sudan and southern Egypt; it can also be found in an isolated area in the Cape Province. Its preferred habitats are dry wadi beds within broad valleys. The ostrich is omnivorous but feeds mainly on plants. The call of the ostrich is described by Cramp and Simmons (1977) with reference to Sauer (1970). A wide range of vocalizations is reported, including hoarse and guttural, sometimes clear calls; hissing; snoring; and booming. During courtship display, males boom day and night, especially in the early morning and the late afternoon. The calls are audible from up to 3 km away (Brown et al. 1983). The vocalizations of chicks are "melodious" contact calls, which can be heard for the first time some days before hatching and are thought to synchronize hatching.

3.1.7 Barred Button Quail

The barred button quail (*Turnix suscitator*; Fig. 2G) is an Oriental species. Its habitat extends from western India to Okinawa and the Lesser Sunda Islands in the East. It is confined to grasslands and fern steppes. Today, this species also seems to inhabit abandoned cleared land that has not yet been reforested. Button quails can be observed from sea level up to an altitude of 1800 m wherever suitable ecological conditions are found. The small birds show a strong sexual dimorphism in body size and plumage coloration, with the females weighing 72.5 g and the males being distinctly smaller, weighing only 45.7 g. Beside this sexual dimorphism, a sex role "reversal" is described, with females courting males and a polyandric mating system probably consisting of a succession of males. Care for the young is probably exclusively paternal (Starck 1989, 1991a,b). No reports exist on the acoustic behavior of the barred button quail. From my own observations, a soft booming is uttered by the females. No other vocalizations are known.

Summarizing this section, the acoustic communications of paleognathous birds and of Turnix can be described as relatively simple, low-frequency calls. Complex calls and duet songs have been reported only in the tinamou. Unfortunately, the hearing range is not known for any of these species.

3.2 External Ear Opening: Position, Ear Covers, Form, and Width of Opening

American Rhea. In the American rhea, the opening of the external auditory meatus faces the lateral side of the head, just behind the corner of the mouth. The ear opening has an oval outline, measuring 30 × 20 mm at maximum extension. A few short, operculate feathers are present in the anterior and posterior auricular tract (pteryla auricularis), providing only an incomplete protective cover for the opening. The operculate feathers are long semibristles or bristles (setae) with a basally thick and elongated rachis (Fig. 3). Few barbs are present on the lateral sides of the rachis. They are denser around the superior umbilicus. A few, scale-like barbules emerge from each barb, in no particular direction and without overlapping other barbules. The scales are restricted to the basal parts of the barbs. The surface of rachis and barbs is smooth, except for some shallow oval impressions found in scanning electron micrographs which are presumed to be impressions of ramogenic cells.

Removal of the skin shows that the external ear opening is completely surrounded by the dorsal superficial fascia of the head. No muscle fibers are attached to the margin of the ear opening (Fig. 18A). After removal of the superficial fascia, the topography of the external ear opening becomes visible in detail. Its occipital margin is formed by the postmeatic membrane (aponeurosis of the depressor mandible muscle). Dorsally, it is bordered by the ventral margin of the squamosal bone and its zygomatic process, whereas the anterior and ventral borders are not clearly defined and are surrounded by loose connective tissue. This tissue separates the external auditory meatus from the kinesis of the quadrate and quadratojugal bar, which are located in close proximity dorsally and ventrally (Fig. 4C).

Crested Tinamou. In the crested tinamou, the external opening of the auditory meatus faces the lateral side of the head as described for the rhea. Relative to the size of the head, the ear opening is small and situated more caudally than in the rhea. Its round shape has a diameter of 4 mm, resulting in an area of 12.6 mm^2. The external ear opening of the crested tinamou is densely covered with operculate feathers of the anterior auricular tract, which bend uniformly in the occipital direction and cover the ear opening as well as the much shorter feathers of the posterior auricular tract.

The barbs of the operculate feathers in tinamous have proximal barbules with a short, flat base and a short pennulum, thus showing all the characteristics of an unmodified proximal barbule. Distal barbules, however, are more or less reduced, except for a few distal barbules in the corners between barbs and rachis. These have ventral teeth and hooklets overlapping the barbules of the adjacent barb. Because of the lack of distal barbules, the feathers do not

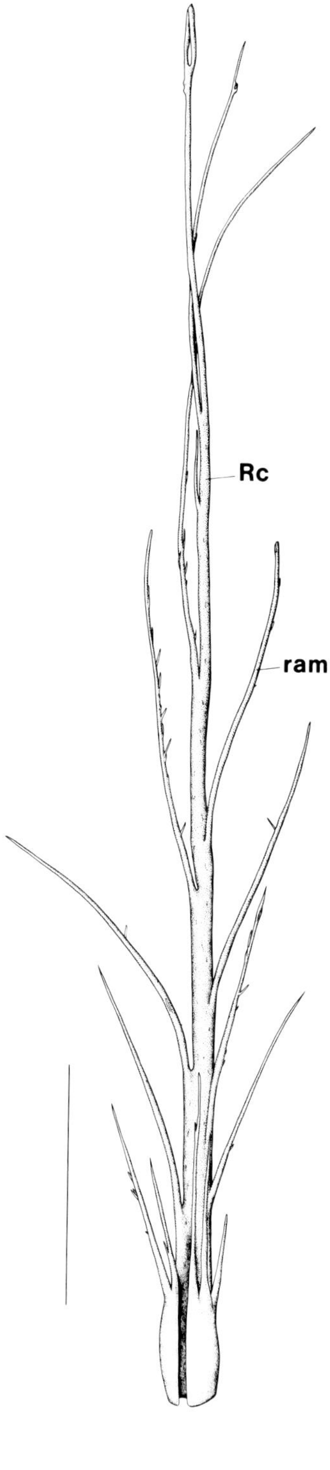

Fig. 3. American rhea, semibristle of the anterior auricular tract. Redrawn from scanning electron micrograph, scale bar = 1 mm

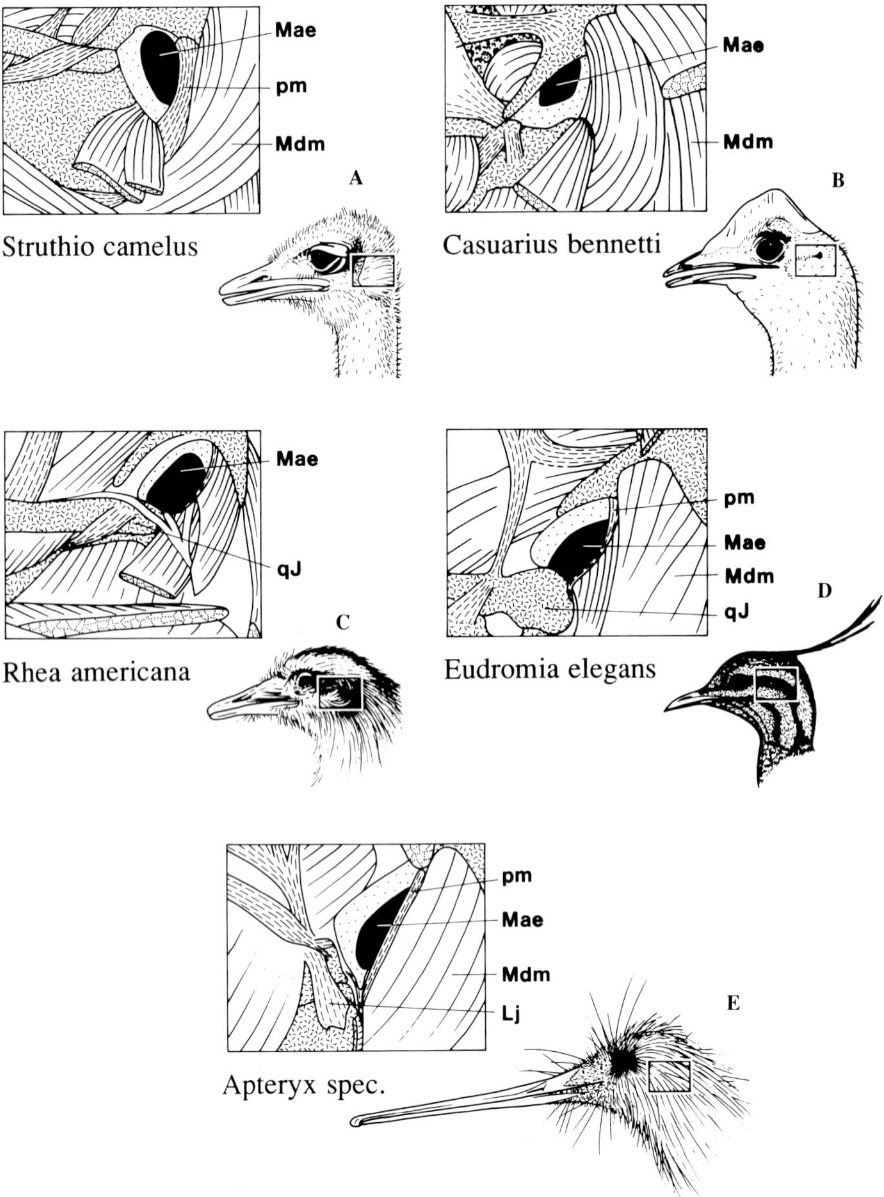

Fig. 4A–E. Topography of the external ear opening after removing skin and superficial fascia. **A** Ostrich. **B** Dwarf cassowary. **C** American rhea. **D** Crested tinamou. **E** Kiwi. Drawn from dissection photographs

constitute a vane; rather, they resemble a lattice (Fig. 5). As in the rhea, the surface shows oval and shallow impressions.

Removal of the skin shows that the intermandibular part of the constrictor collis muscle (m. constr. coll. pars intermandibularis) is attached to the anter-

Fig. 5. Crested tinamou, scanning electron micrograph of an operculate feather of the anterior auricular tract. Scale bar = 0.2 mm

oventral margin of the ear opening (Fig. 18B). The next deeper layer shows the same topography as described for the American rhea (Fig. 4D).

Other Tinamou Species. Material from other tinamou species was available for comparison. The ear opening of the white-throated tinamou (*Tinamous guttatus*), which is one of the larger forest tinamous, resembles that of the crested tinamou. It has a small and round opening 4 mm in diameter. Operculate feathers of the anterior auricular tract cover the ear opening as described for the crested tinamou. In all other tinamou species studied (*Crypturellus tataupa*, *C. soui*, *C. obsoletus*, and *Nothoprocta pentlandii*), the pattern of auricular feathers is similar to that just described.

Emu. The external ear opening of the emu faces the latero-occipital direction. This is comparable to that of the ostrich (see following). The external ear opening of the emu has a highly oval outline measuring 23 mm × 17 mm. The integument of the lateral and occipital regions of the head is feathered, though sparsely, and has a yellow and blue coloration. The skin of the external auditory meatus completely lacks feathers and shows a deep brown pigmentation.

In the emu, the feathers of the lateral head surface cannot be assigned to any feather tract. However, feathers anterior to the ear opening are longer than others, directed backwards, and obviously function as protective operculate feathers, although the cover is incomplete. The feathers posterior to the ear opening are directed laterally or caudally and do not contribute to the opercularization of the ear opening.

Scanning electron microscopy did not reveal differences between the feathers surrounding the ear opening and the feathers of the head. All have a strong rachis, relatively short barbs, and reduced barbules. They are semi-bristles, although less modified than in the rhea.

Cassowary. The cassowary has a mostly featherless head. The skin of the head is highly modified by extended subcutaneous sinusoids, which can become distended with rising blood pressure. Filling of the venous sinusoids results in an inflation of the head and neck regions, and probably in a color change as well. Behavioral observations document inflation in the context of booming display posture (Coates 1985). All integumentary structures of the cassowary's head are modified by this subcutaneous organ. Modifications of the subcutis can already be found in hatchlings, in whom the skin is densely vascularized. In the adult, the capillaries of the stratum laxum dermi are modified into large cavernous sinusoids, which extend through all of the stratum laxum and modify the opening of the external auditory meatus.

The external ear opening lies ca. 25 mm behind the posterior corner of the mouth. It measures ca. 12.5 mm in diameter, is irregularly shaped, and small compared to that of other paleognaths. A prominent, horizontal cutaneous rim is present just between the posterior edge of the mouth and the anterior margin of the ear opening. In live birds, this cutaneous fold is a brilliant red and contrasts with the dull brown coloration of the head. In macroscopic dissection, the rim is tightly fixed to the superficial fascia and is densely supplied with sinusoids. Its histological structure suggests that it can be inflated in the same way as the other regions of the skin. Thus, inflation may affect the shape and the width of the external ear opening.

Only a few true bristles having a thick, elongated shaft and no barbs or barbules are present on the head of the cassowary. The bristles are sparse and cannot be assigned to any feather tract. The external ear opening is unprotected. The bristles around the ear opening do not differ from other bristles. Removal of the skin and superficial fascia reveals the same topography of the external ear opening as that found in the other species (Fig. 4B).

Kiwi. The external ear opening of the kiwi is situated on the lateral side of the head, half-way between the corner of the mouth and the occiput. It is circular, with a diameter of 6.5 mm and an area of 33.18 mm^2 when open. It is densely covered by feathers. The structure of these feathers resembles those of the other feathers of the head. Like the cassowary, the kiwi has no distinct feather tracts and it is not possible to describe an anterior or posterior auricular tract. The feathers of the kiwi are peculiar in that their proximal and distal barbules are elongated, forming long "hair-like" filaments that do not hook into each other. Thus, the feathers appear plumaceous although they are actually contour feathers. The feathers anterior to the ear opening are deeply inserted into the fascia, comparable to sinus feathers. They probably serve a tactile function, as those around the mouth opening. Scanning electron microscopy reveals pointed scales on the surface of the feathers. These scales might represent reduced barbicels and should not be confused with the reduced barbules described in the rhea.

Removal of the skin shows some fibers of the constrictor collis muscle close to the ear opening, but they do not insert into it. The topography of the next deeper layer is the same as that just described for the other species (Fig. 4E).

Ostrich. The external ear of the ostrich opens to the lateral and occipital sides of the head in a caudal and ventral position. It has a backward orientation. Its outline roughly resembles a rounded triangle, with curved sides and rounded corners. The opening measures ca. 40 × 45 mm. A skin fold operculates the anterior part of the external ear opening. The superficial fascia inserts into this skin fold, but there are no muscle fibers attached to it. Here again, the deeper topography (Fig. 4A) resembles that described in the American rhea.

Operculate feathers of the anterior and posterior auricular tract cover the opening. The anterior and posterior auricular feathers differ in length and orientation. Those of the anterior tract are long, pointing to the occipital region and covering the external ear opening. The feathers of the posterior tract are shorter; they also point to the occipital direction and consequently do not contribute to a covering of the ear opening. The operculate feathers have a semibristle structure similar to those of the American rhea. The main difference between the two species is that the scale-like barbules are more numerous in the ostrich, especially in the apical regions of its barbs.

Barred Button Quail. The ear of the barred button quail opens wide on the lateral surface of the head just behind the posterior corner of the mouth. It is sparsely covered by stiff auricular feathers. The ear opening is rather large for this small bird, measuring 2–4 mm.

The feathers of the anterior auricular tract reach the posterior margin of the ear opening. They have a long rachis but a narrow vane, thus constituting only an incomplete fence-like cover of the ear opening. Feathers of the posterior auricular tract are short and do not cover the ear opening.

3.3 External Auditory Meatus (Meatus acusticus externus): Position and Course, Bordering Structures, and Relation to Fascia and Muscles

Despite the variety in types of external ear opening, the external auditory meatus shows only minor variations within the paleognathous birds. Its description is therefore based on the American rhea.

The external auditory meatus is a short and wide tube. From the lateral side to the tympanic membrane, it first bends ventrally before it turns dorsally to reach the tympanic membrane from the ventral side. The external auditory meatus is funnel-shaped, with its largest cross section at the tympanic membrane (Figs. 6–17).

In all species, the inner surface of the external auditory meatus is covered by bare skin, with no feathers. Towards the tympanic membrane, the skin gradually becomes thinner and continues into the external layer of the tympanic membrane. There are no differences within the group of paleognathous birds

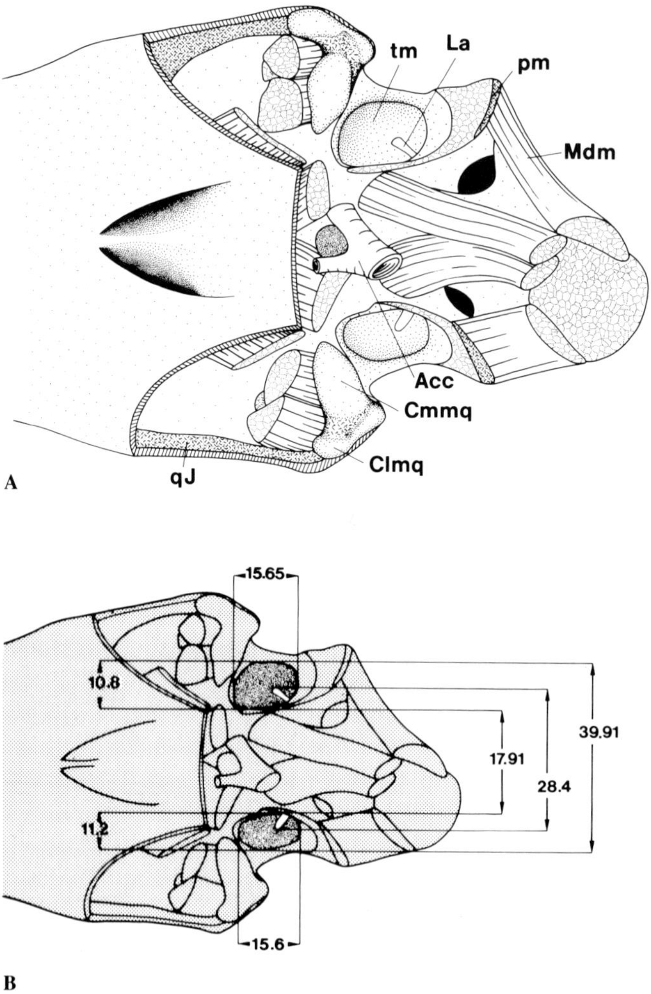

Fig. 6A,B. American rhea, adult. **A** Dissection of the external auditory meatus and the tympanic membranes, palatal view. **B** Topographic morphometry of the tympanic membrane

except for the specific modifications of the ear opening described in the previous paragraph. The auditory meatus of the barred button quail is short and leads directly, without bending, to the tympanic membrane, which in this species is oriented to the lateral side.

As in all other birds, the membrana postmeatica is the ventral and occipital wall of the auditory meatus (Fig. 4A–E). It represents the aponeurosis of the depressor mandibule muscle. Anteriorly, the posterior edge of the quadrate bone contributes to the wall of the auditory meatus; however, loose connective tissue separates both structures. This might reduce the impact of mouth opening and kinesis of the quadrate bone on the cross-sectional area of the external auditory meatus. The superficial fascia, which in some species contains muscle fibers, enters at the margin of the external ear opening (Fig. 18A,B).

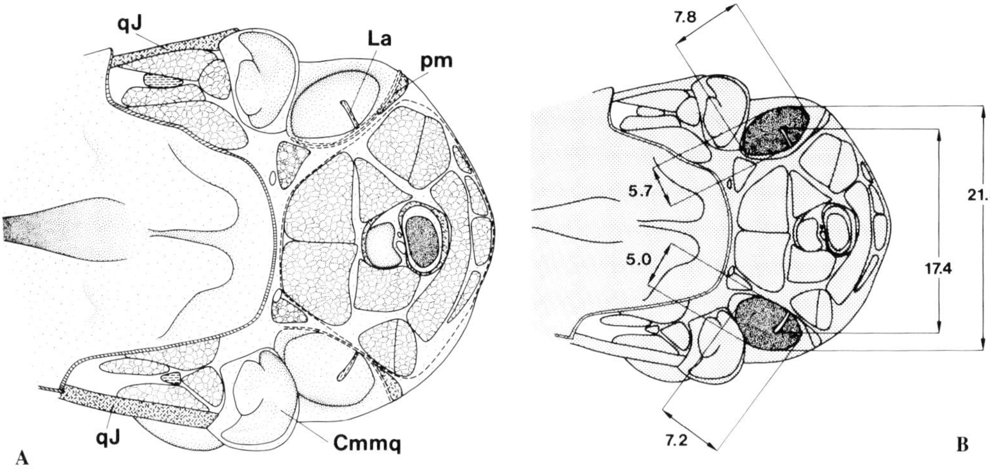

Fig. 7A,B. Crested tinamou, adult. **A** Dissection of the external auditory meatus and the tympanic membranes, palatal view. **B** Topographic morphometry of the tympanic membrane

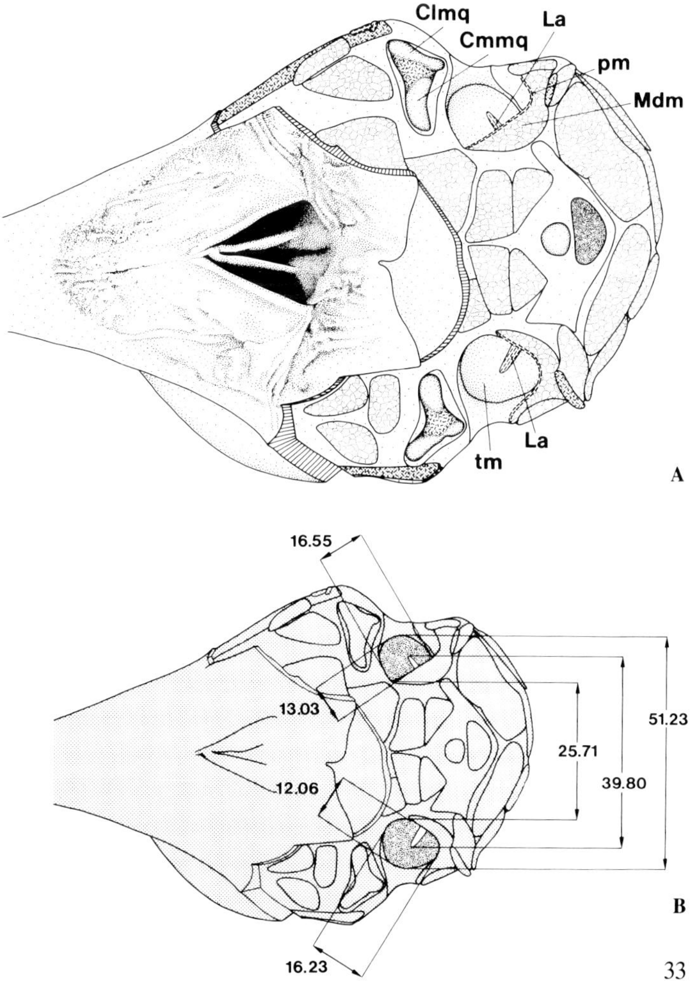

Fig. 8A,B. Dwarf cassowary, adult. **A** Dissection of the external auditory meatus and the tympanic membranes, palatal view. **B** Topographic morphometry of the tympanic membrane

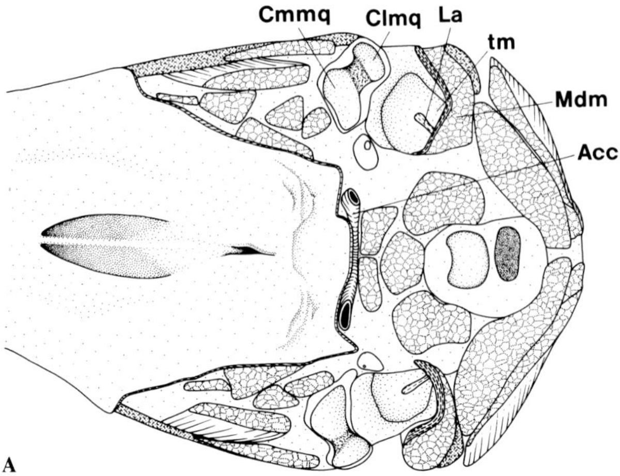

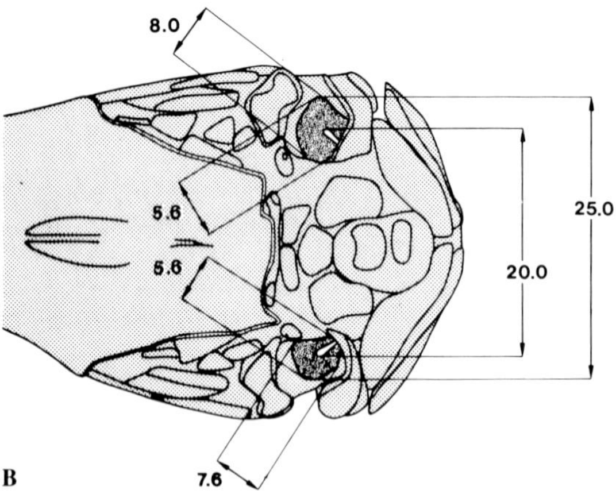

Fig. 9A,B. Kiwi, adult. **A** Dissection of the external auditory meatus and the tympanic membranes, palatal view. **B** Topographic morphometry of the tympanic membrane

Comparison. The external auditory meatus is passively protected by the operculate feathers in the tinamous, the ostrich, and the kiwi. The auricular feathers of the American rhea, the emu, and the barred button quail are not suitable for protection. The cassowary also has unprotected ear openings, which, however, are smaller than in the other species. Swelling of the skin might close the ear opening, but this is obviously merely a side effect of booming display. It is highly improbable that the feathers around the ear have any effect on the hearing capabilities of the birds as described for other species (Iljitschew 1967, 1968, 1974). In none of the species investigated here is there a specialized opercular mechanism comparable in development to that of diving birds, e.g., Alcidae (Kartaschew and Iljitschew 1964) or Spheniscidae (Anisimov and Iljitschew 1980a). Sebaceous skin glands of the auditory meatus (glandulae auriculares), which have been described in galliform birds (Plate 1918; Lucas and Stettenheim 1972), are not found in palaeognaths or turnix.

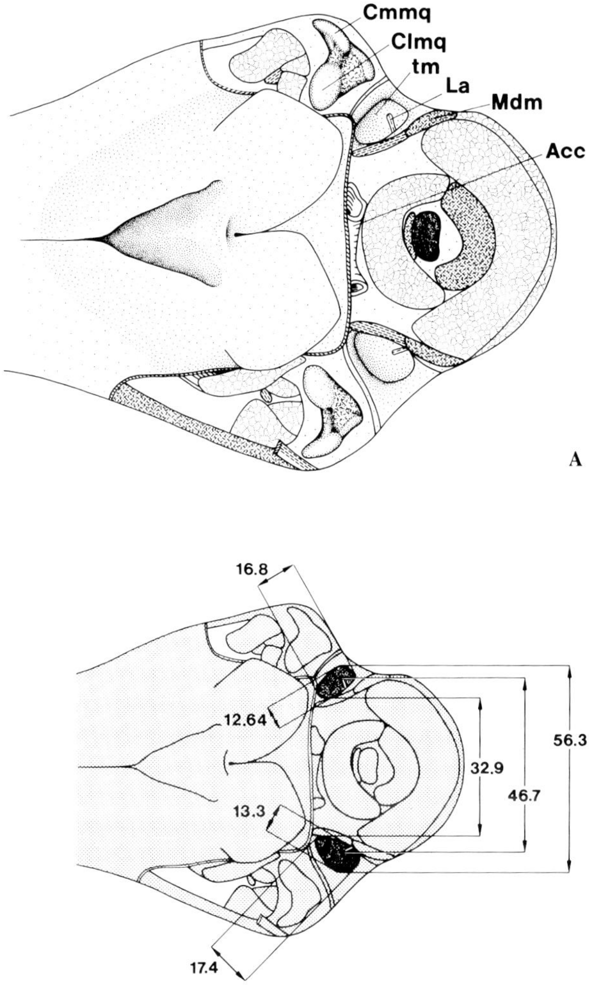

Fig. 10A,B. Ostrich, adult. **A** Dissection of the external auditory meatus and the tympanic membranes, palatal view. **B** Topographic morphometry of the tympanic membrane

Previous descriptions of "cutaneous ear folds" related to auricular glands in the American rhea (Schwalbe 1890) cannot be confirmed.

3.4 Osteology of the Tympanic Region

Several cranial elements contribute to the bony enclosure of the tympanic region. Depending on age, size, and systematic position, the bony elements involved differ not only among species but also among the various developmental stages within a species. The tendency of connective tissue to ossify with age makes any description difficult since the bony enclosure of the tympanic

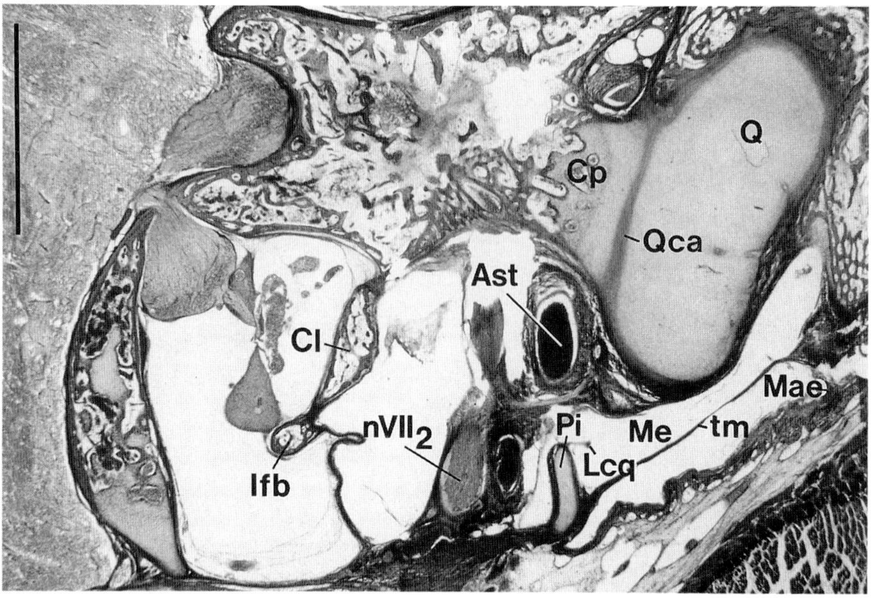

Fig. 11. Emu. Cross section through the head of a hatchling. Microphotograph of the otic region, celloidin, serial section #597, 40 μm, azan, scale bar = 1 mm

region also includes ossifications that cannot be assigned to distinct bones in terms of comparative morphology.

American Rhea. The otic capsule and canalicular capsule constitute the dorsal wall of the tympanic cavity. They ossify from the prootic bone in the rostromedial and dorsal parts and from the exoccipital bone in its posterior parts (Fig. 19). The prootic bone and the exoccipital bone meet at the "interfenestral bar" (promontorium). The supraoccipital bone also contributes to the ossification of the canalicular part but has no direct share in the ossification of the tympanic region in the hatchling. In later stages of development, the posterior tympanic recess pneumatizes the supraoccipital bone and establishes a connection between the supraoccipital bone and the tympanic cavity. The occipital wall of the tympanic cavity is built from parts of the metotic cartilage (cartilago metotica). Its ossification extends from the exoccipital bone. In adult birds, this structure is described as the tympanic wing of the exoccipital bone (ala tympanica exoccipitalae, Fig. 20). Ventromedially, the tympanic cavity is closed by the basitemporal plate. This complex comprises several dermal bones[1]

[1] In Anglo-American morphology, all bones of direct membraneous ossification are described as *dermal* bones, regardless of their position. In contrast, European morphology distinguishes between position and histogenesis. *Dermal* refers to the phylogeny and the origin of the of the bone from the exoskeleton of ancestral vertebrates. *Desmal* refers to the histogenesis of a bone as membraneous ossification. The terms are not synonymous, although there is a broad overlapping in meaning. In this paper, I am using the Anglo-American terminology, describing all membraneous bones as dermal regardless of their position.

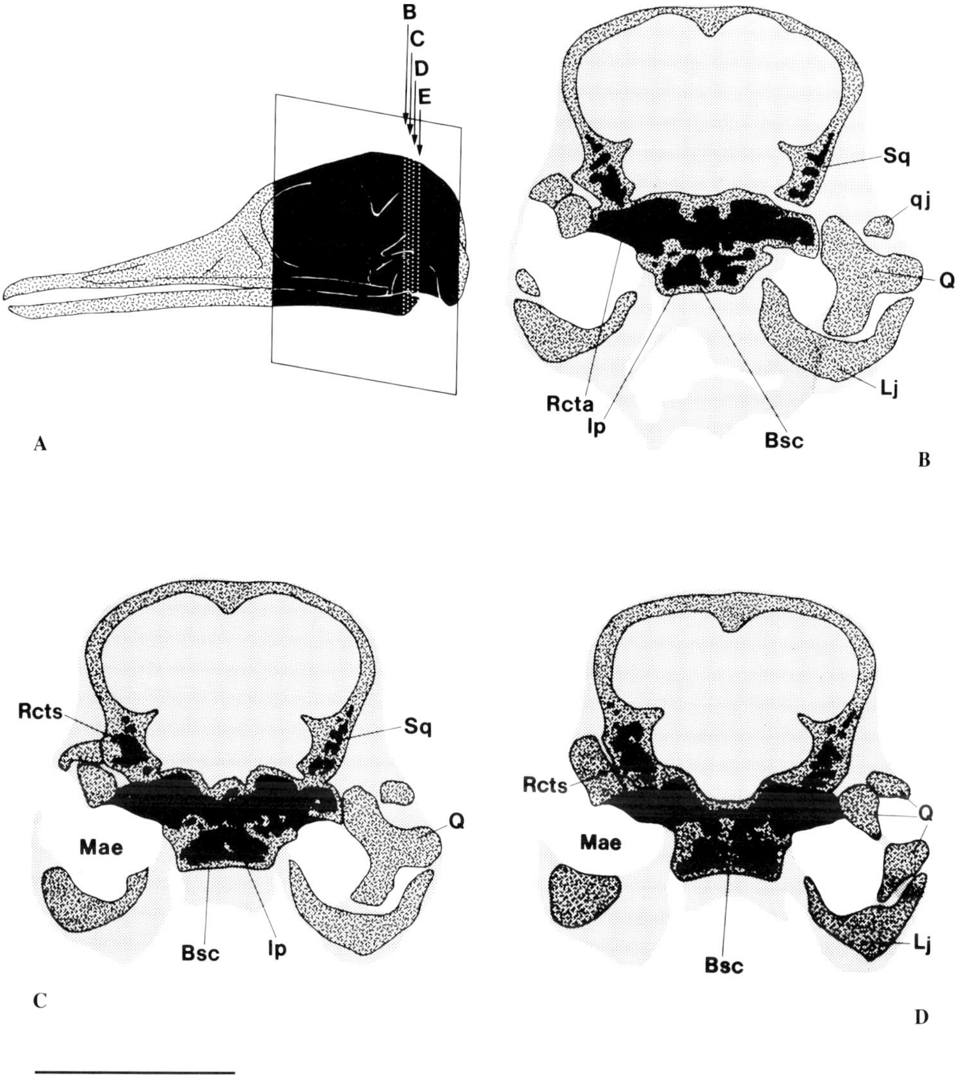

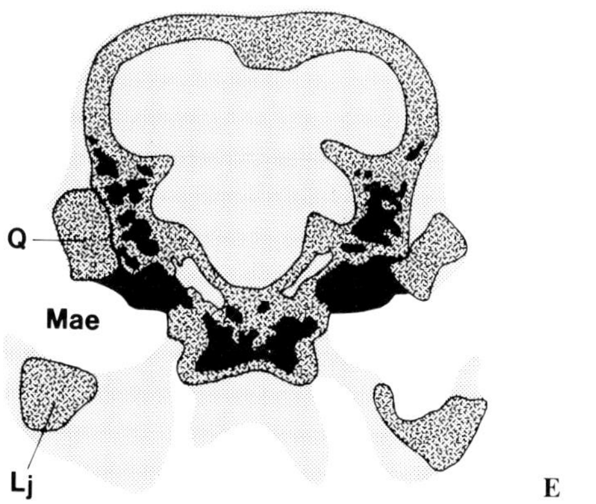

Fig. 12A–E. American rhea, adult. X-ray computed tomography of the otic region. **A** Scanned region (*black*) and inclination of CT scans. Position of individual scans (**B–E**) is indicated by *white dotted lines* and *arrows*. **B** Scan no. 50, **C** no. 51, and **D** no. 53 consist of successive scans through the interaural pathway. **E** Scan no. 56 through the tympanic cavity at the level of the oval window. **A–E** Scale bar = 50 mm

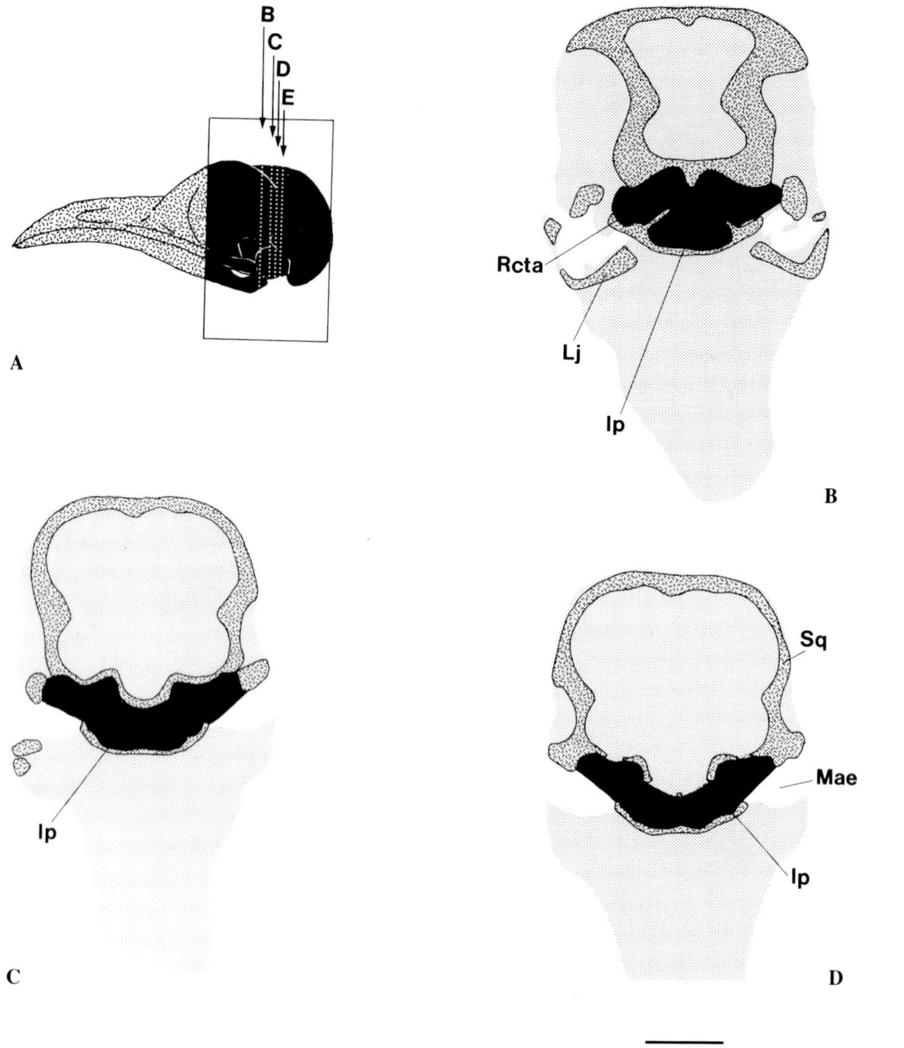

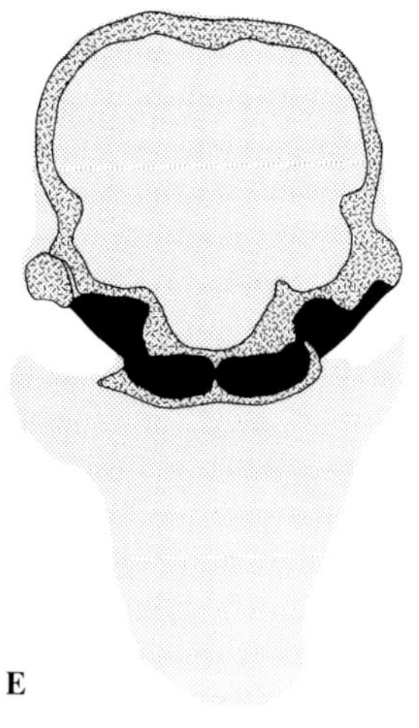

Fig. 13A–E. Crested tinamou, adult. X-ray computed tomography of the otic region. **A** Scanned region (*black*) and inclination of CT scans. Position of individual scans (**B–E**) is indicated by *white dotted lines* and *arrows*. **B** Scan no. 12, **C** no. 14, and **D** no. 15, consist of successive scans through the interaural pathway. **E** Scan no. 16 through the tympanic cavity at the level of the oval window. **A–E** Scale bar = 10 mm

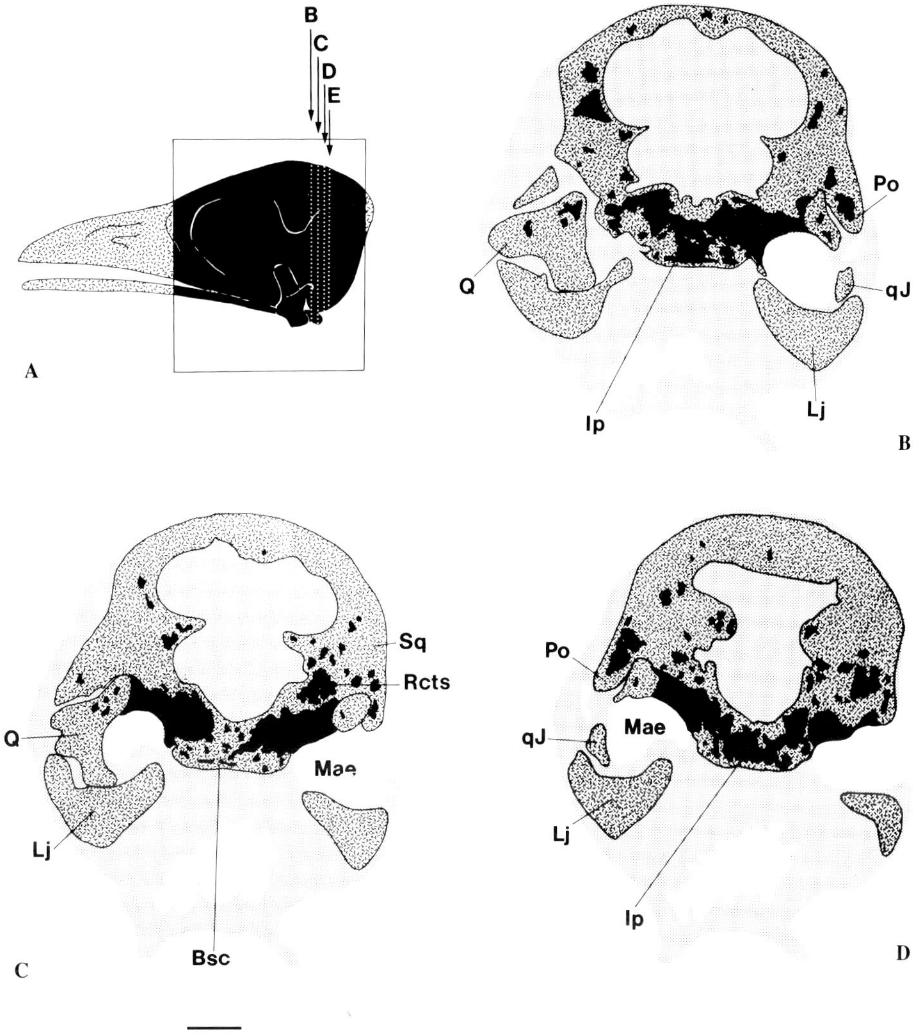

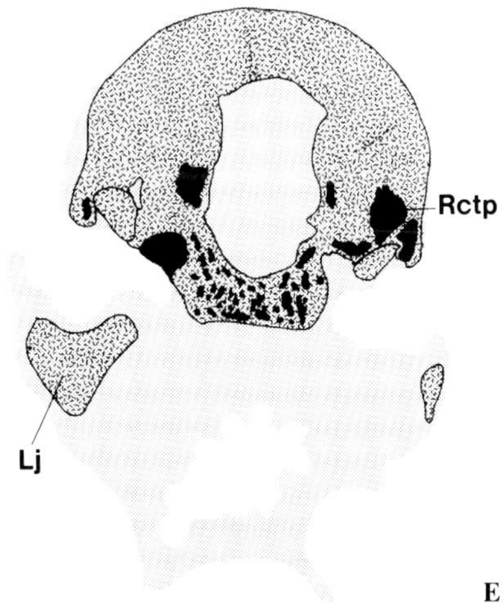

Fig. 14A–E. Emu, adult. X-ray computed tomography of the otic region. **A** Scanned region (*black*) and inclination of CT scans. Position of individual scans (**B–E**) is indicated by *white dotted lines* and *arrows*. **B** Scan no. 58, **C** no. 61, and **D** no. 63 consist of successive scans through the interaural pathway. **E** Scan no. 65 through the tympanic cavity at the level of the oval window. **A–E** Scale bar = 50 mm

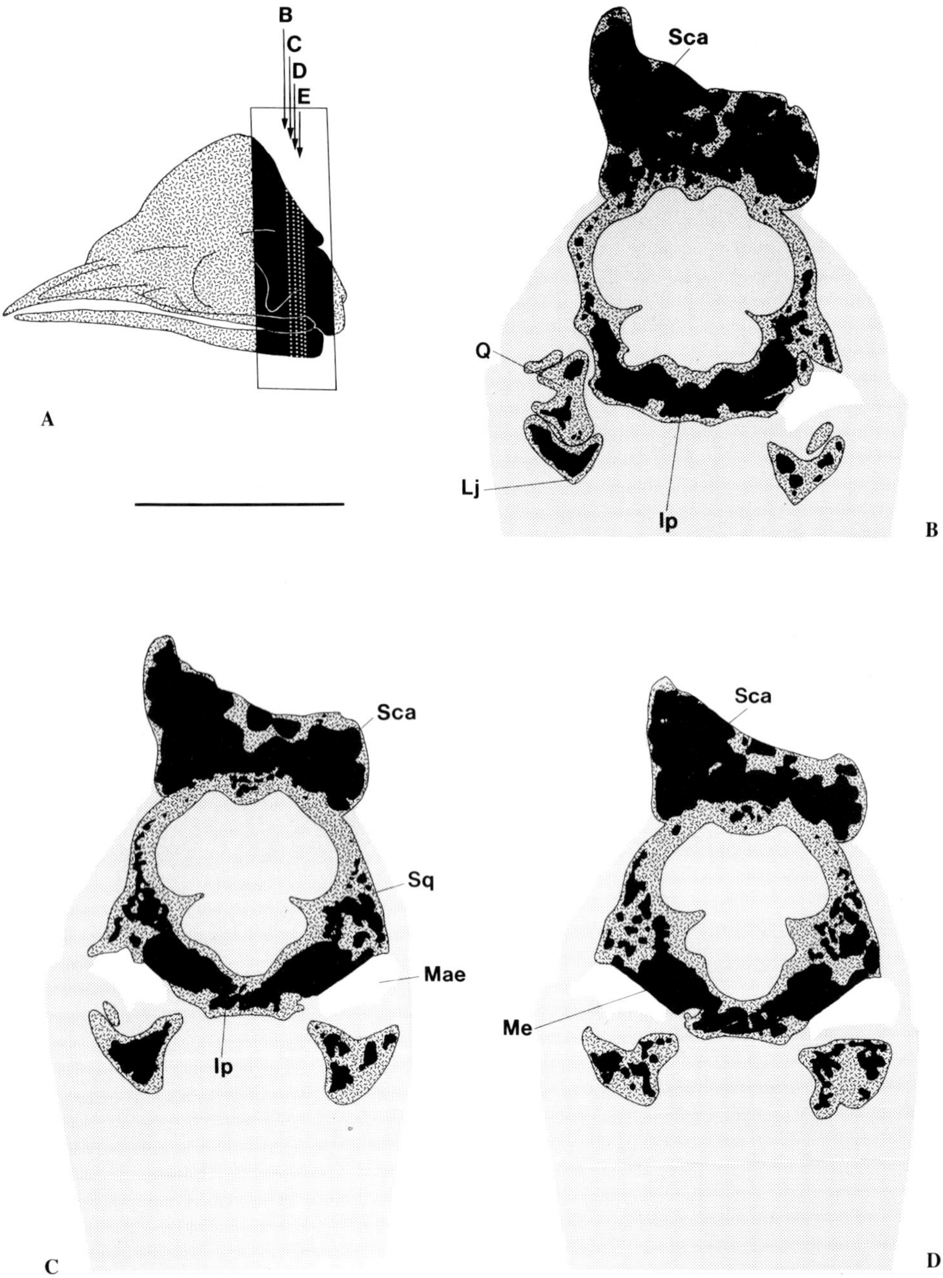

Fig. 15A–E. Cassowary, adult. X-ray computed tomography of the otic region. **A** Scanned region (*black*) and inclination of CT scans. Position of individual scans (**B–E**) is indicated by *white dotted lines* and *arrows*. **B** Scan no. 18, **C** no. 21, and **D** no. 23 consist of successive scans through the interaural pathway. **E** Scan no. 25 through the tympanic cavity at the levle of the oval window. **A–E** Scale bar = 50 mm

Fig. 15A–E. *Continued*

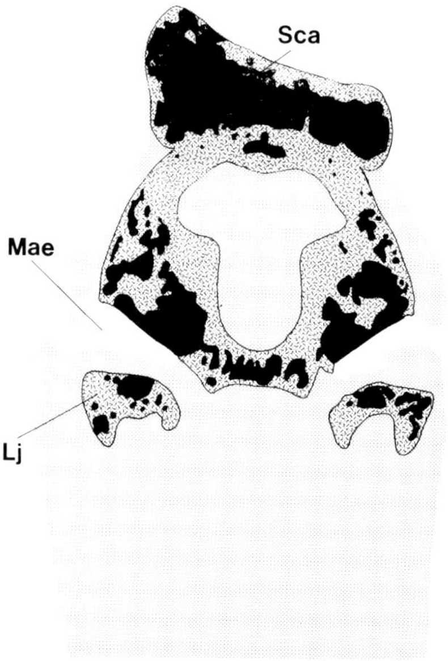

(the parasphenoid, basiparaspenoid, rostroparasphenoid, and alaparasphenoid) as well as enchondral ossifications (basisphenoid). All ossifications fuse early in embryogenesis to form the basitemporal plate. Rostrally, the lateral wing of the parasphenoid builds a more or less perpendicular bony wall as anterior closure. Between this structure and the ventral wing of the parasphenoid, a bony tunnel is formed, enclosing the pharyngotympanic tube (Eustachian tube). Dorsomedially, the roof of the tympanic cavity consists of the dorsal surface of the basisphenoidal bone, the internal surface of the pleurosphenoid, and the lateral wall of the prootic bone. Dorsolaterally, the squamosal bone closes the tympanic cavity. The otic process of the quadrate reaches into the tympanic cavity, thus contributing to the dorsoanterior wall of the middle ear cavity (see 3.9).

Crested Tinamou. The crested tinamou resembles the American rhea in many respects (Fig. 21). Differences are present in the tympanic wing of the exoccipital bone (cartilago metotica of the hatchlings), which is distinctly smaller and narrower than in the rhea. The same is observed in the basisphenoid complex, which does not extend as far laterally as in the rhea.

Emu. The same pattern of ossification surrounds the tympanic cavity in the emu (Fig. 22). The metotic cartilage of the hatchling and the tympanic wing of the adult bird are not as large as in the rhea, but also form a wall caudal to the tympanic cavity. The squamosal bone has a long zygomatic process. Posteriorly, the squamosal overlaps the tympanic wing of the exoccipital. In the hatchling, the lateral and the ventral wings of the basiparasphenoid form a dorsal and a medioventral closure of the tympanic cavity. As described in the other species,

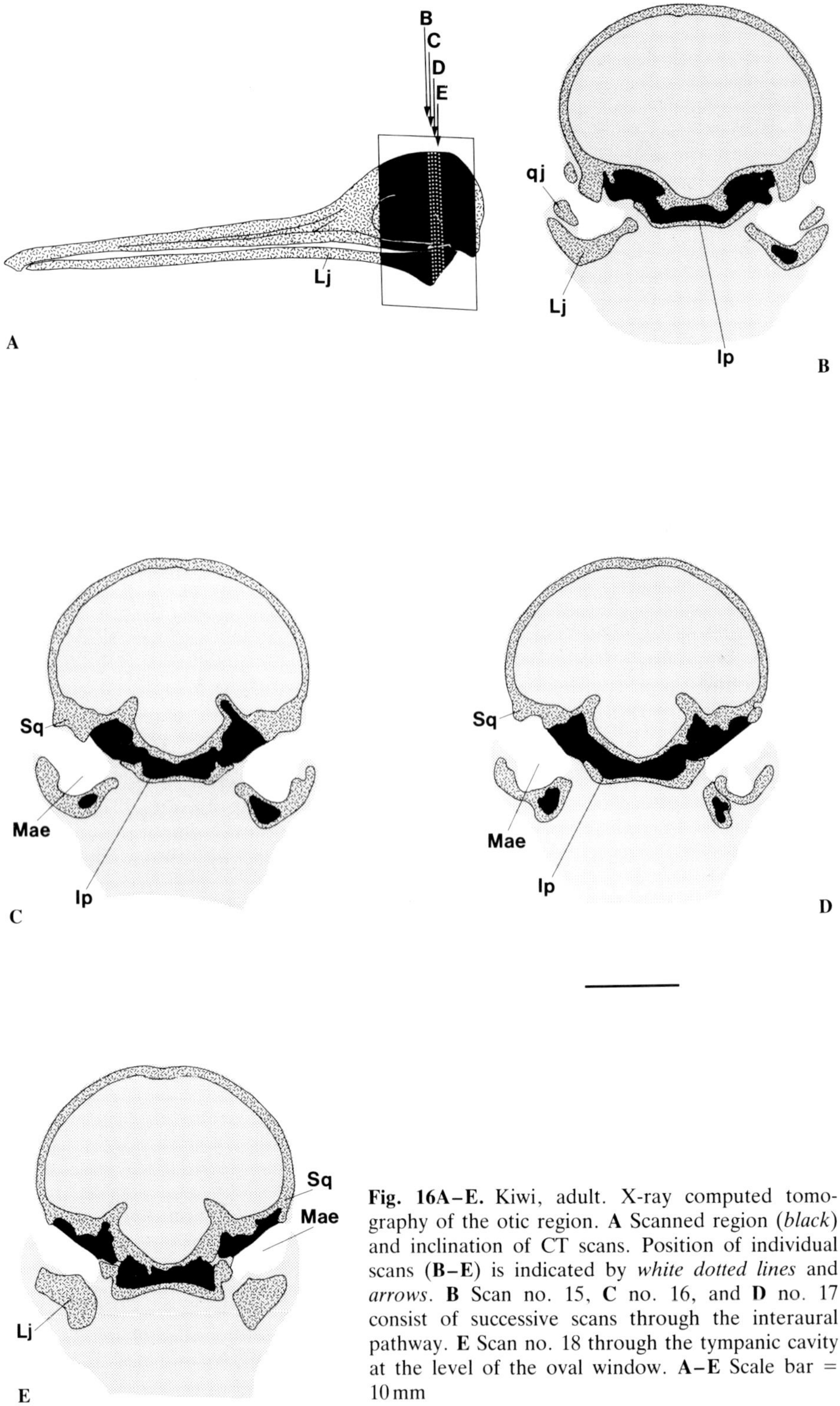

Fig. 16A–E. Kiwi, adult. X-ray computed tomography of the otic region. **A** Scanned region (*black*) and inclination of CT scans. Position of individual scans (**B–E**) is indicated by *white dotted lines* and *arrows*. **B** Scan no. 15, **C** no. 16, and **D** no. 17 consist of successive scans through the interaural pathway. **E** Scan no. 18 through the tympanic cavity at the level of the oval window. **A–E** Scale bar = 10 mm

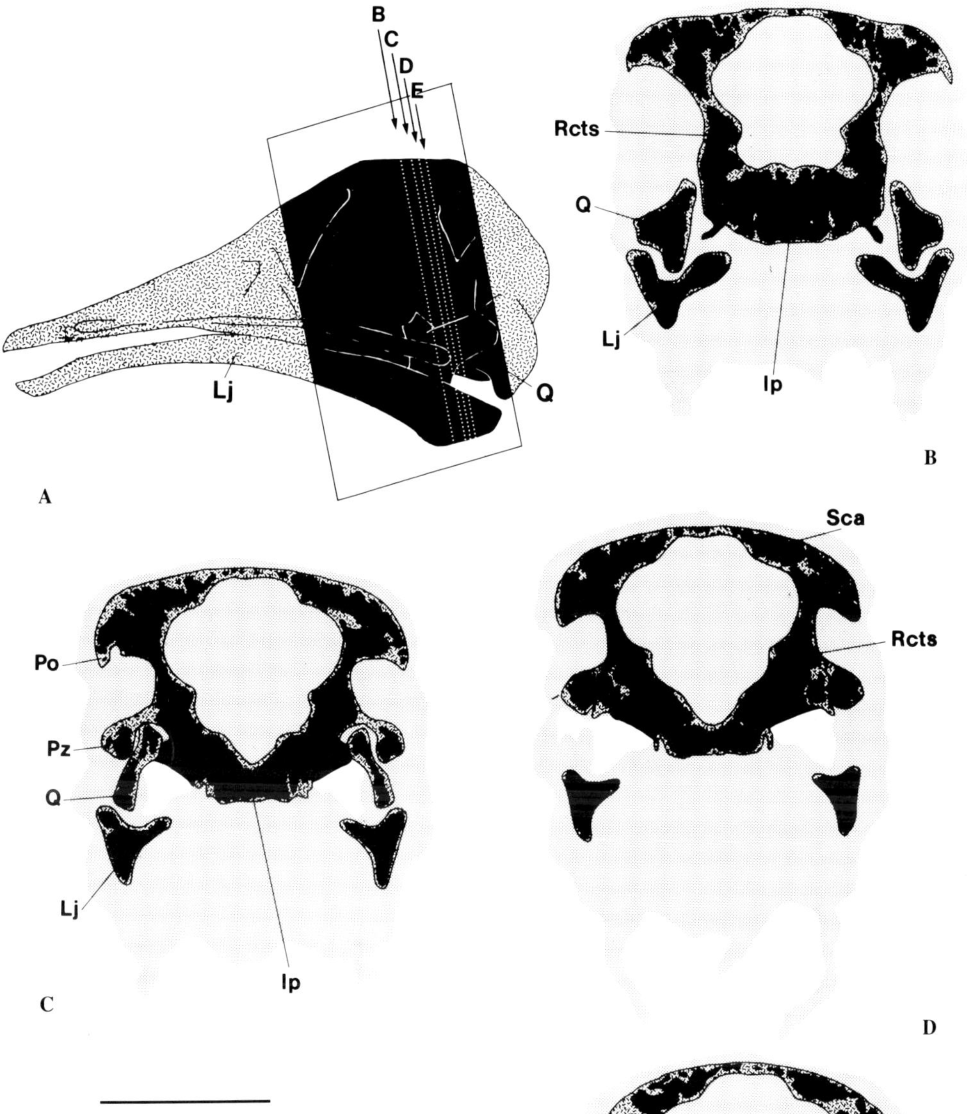

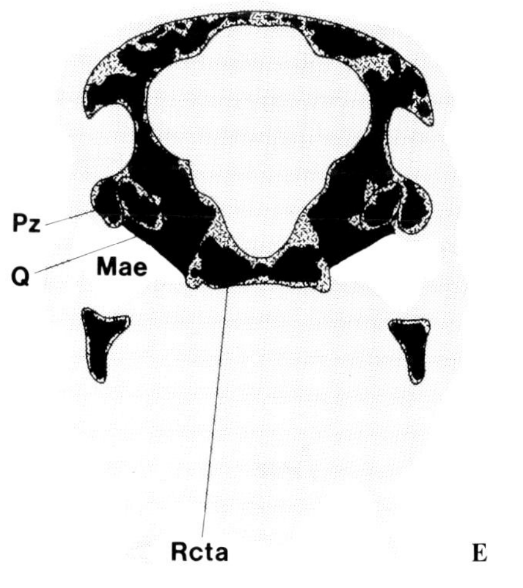

Fig. 17A–E. Ostrich, adult. X-ray computed tomography of the otic region. **A** Scanned region (*black*) and inclination of CT scans. Position of individual scans (B–E) is indicated by *white dotted lines* and *arrows*. **B** Scan no. 47, **C** no. 51, and **D** no. 53 consist of successive scans through the interaural pathway. **E** Scan no. 55 through the tympanic cavity at the level of the oval window. **A–E** Scale bar = 50 mm

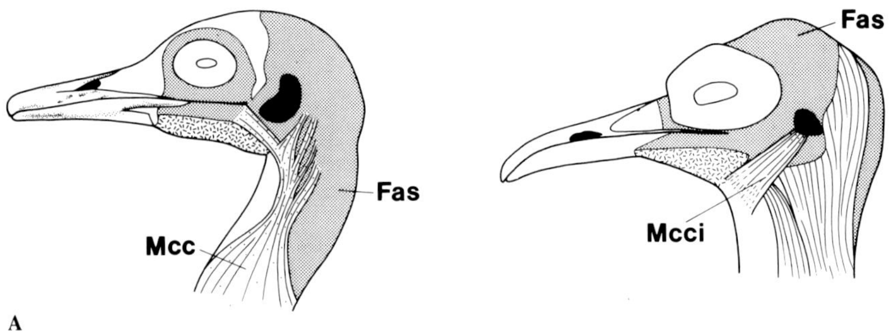

Fig. 18A,B. Superficial fascia and subcutaneous muscles of the head. **A** American rhea. **B** Crested tinamou. Drawn from dissection photographs

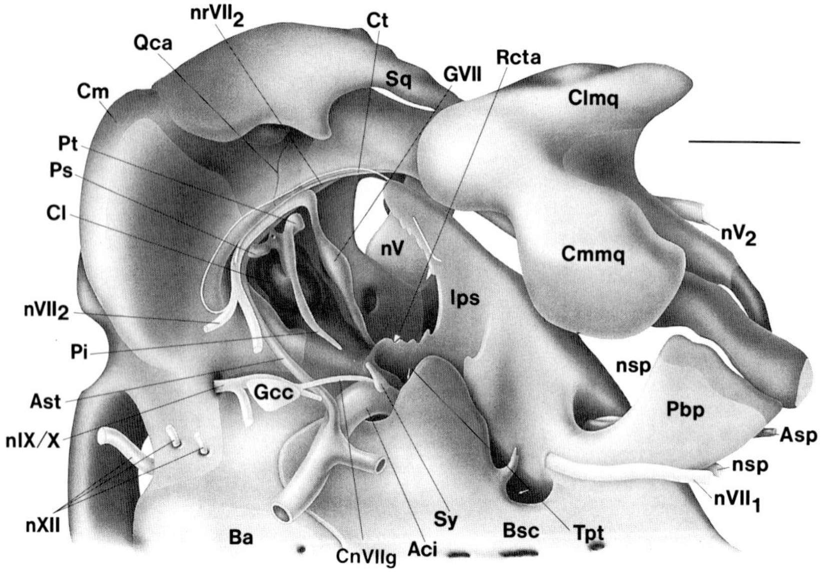

Fig. 19. American rhea. Plate reconstruction of the right middle ear and basicranium of a hatchling. Section thickness for reconstruction, 80 μm. *Right*, rostral; *left*, occipital; *up*, right; ventral view; scale bar = 1 mm

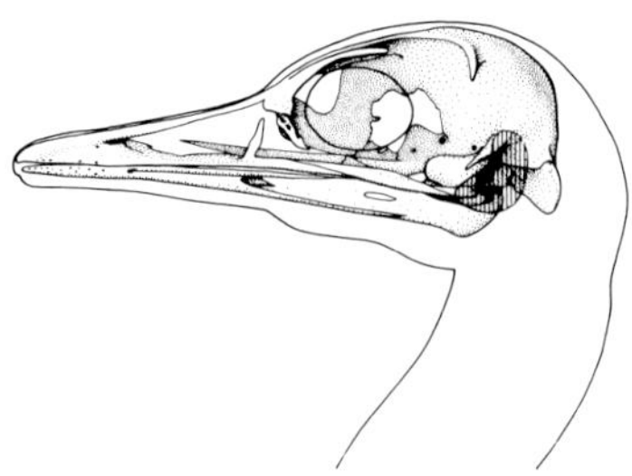

Fig. 20. American rhea, adult. Lateral view of the skull, drawn from skull

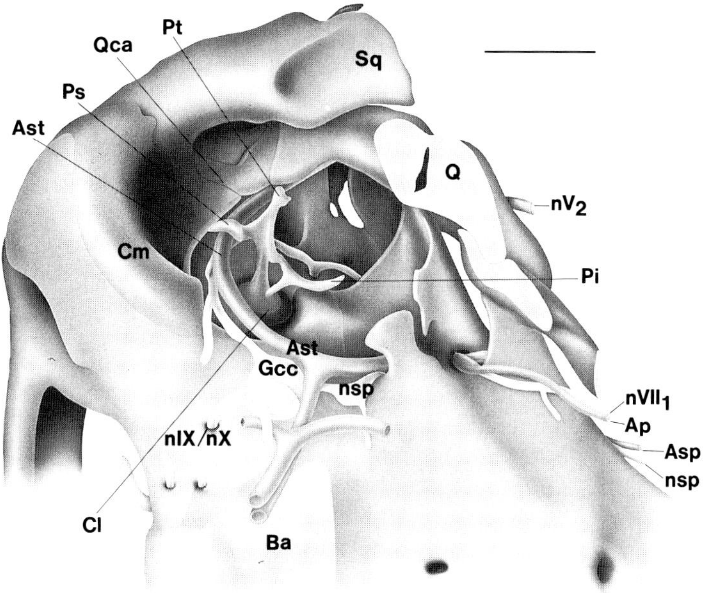

Fig. 21. Crested tinamou. Plate reconstruction of the right middle ear and basicranium of a hatchling. Section thickness for reconstruction, 60 μm. *Right*, rostral; *left*, occipital; *up*, right; ventral view; scale bar = 1 mm

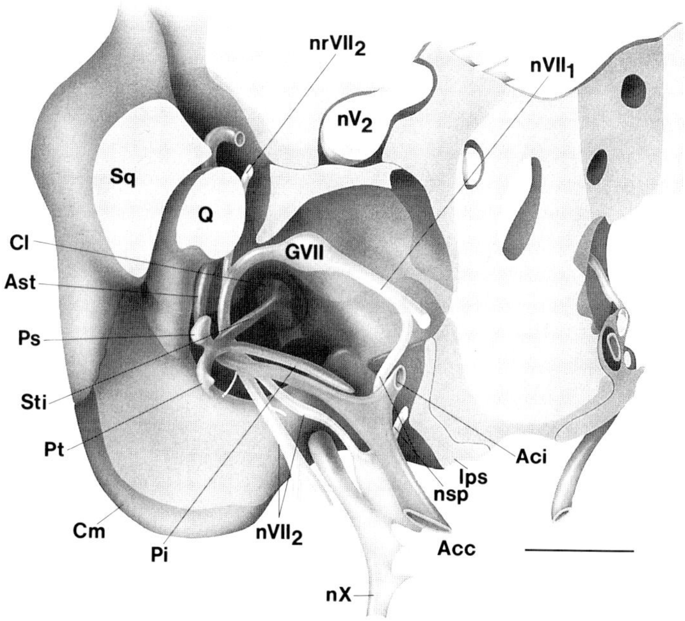

Fig. 22. Emu. Plate reconstruction of the right middle ear and basicranium of a hatchling. Section thickness for reconstruction, 80 μm. *Right*, rostral; *left*, occipital; *up*, right; fronto-lateral view; scale bar = 1 mm

the prootic bone and the exoccipital bone form the medial wall of the tympanic cavity; the contribution of the basisphenoid can be seen in Fig. 22.

Cassowary. The southern cassowary (*Casuarius casuarius*) resembles the other species described here. However, the tympanic wing of the exoccipital bone is rather small and oriented perpendicularly, thus forming the occipital closure of the tympanic cavity (Figs. 23, 24). The embryonic metotic cartilage persists for some time after hatching. In hatchlings, a large metotic cavity is present between the otic capsule and the metotic cartilage. This cavity later becomes integrated into the pneumatization of the skull of the adult bird, and then cannot be distinguished anatomically. As in the other species, the ventromedial

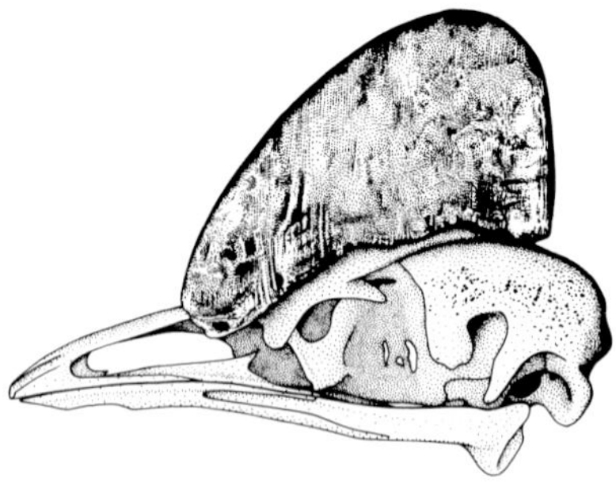

Fig. 23. Dwarf cassowary. Lateral view of the skull

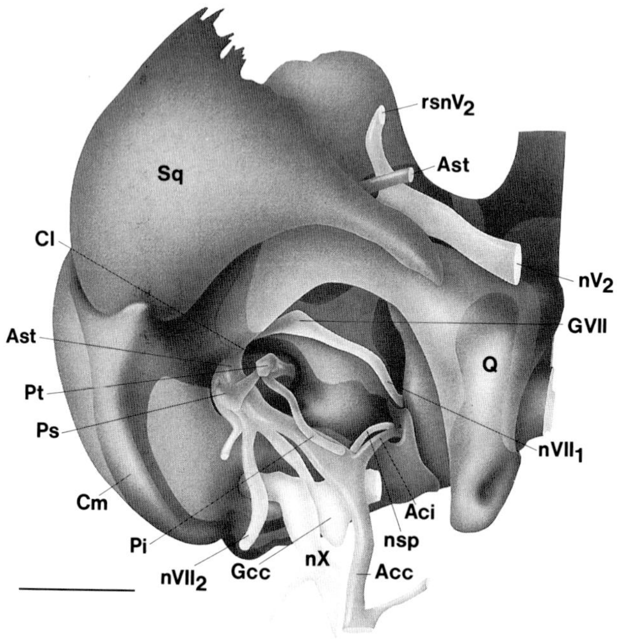

Fig. 24. Southern cassowary. Plate reconstruction of the right middle ear and basicranium of a hatchling. Section thickness for reconstruction, 80 μm. *Right*, rostral; *left*, occipital; *up*, right; frontolateral view; scale bar = 1 mm

wall of the tympanic cavity is formed by the basitemporal plate, which contributes to the anterior wall of the tympanic cavity, a more or less vertical wall. The anterior tympanic recess intrudes into the basitemporal plate and pneumatizes much of its parts (see following). The tympanic entrance of the anterior tympanic recess into the basitemporal complex is funnel-shaped and occupies most of the tympanic cavity. In adult specimens of *C. casuarius* and *C. bennetti*, a bony septum divides the tympanic cavity into a posterior part, with the perilymphatic and vestibular windows, and an anterior part, with the openings of the anterior tympanic recess, the superior tympanic recess, and the pharyngotympanic tube. In the hatchling, this bony ridge has not yet developed.

Kiwi. A large ventral extension of the tympanic wing of the exoccipital bone is a peculiarity of the adult kiwi skull (Figs. 25, 26). In the hatchling, the metotic cartilage forms a dorsal process reaching rostrally to the lateral side of the squamosal. In the adult bird, this process is fused to the neurocranium. The tympanic wing of the exoccipital forms a concave wall to the occipital side of the tympanic cavity. The tympanic membrane is fixed to its ventral margin (see following). The vestibular and the perilymphatic windows are deeply situated in an occipital corner of the tympanic cavity. In adult birds, they are close to the metotic cavity (cavitas metotica). From the metotic cavity, the posterior tympanic recess emerges to pneumatize the exoccipital and supraoccipital bones. The squamosal bone contributes to the dorsal wall of the tympanic cavity. In

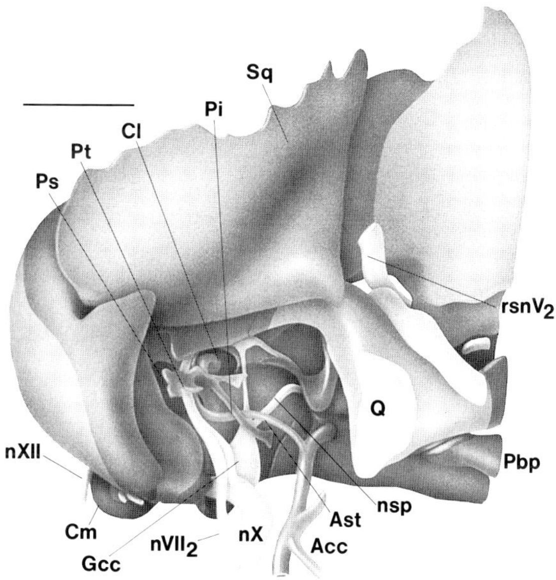

Fig. 25. Kiwi. Plate reconstruction of the right middle ear and basicranium of a hatchling. Section thickness for reconstruction, $80 \mu m$. *Right*, rostral; *left*, occipital; *up*, right; lateral view; scale bar = 1 mm

Fig. 26. Kiwi, adult. Lateral view of the skull

hatchlings, it has only one window for the superior tympanic recess; in adults, it has many.

Ostrich. The ostrich shows no major differences from the species just described (Figs. 27, 28). However, in the adult bird, the tympanic wing of the exoccipital bone is rather small in relation to the large skull.

In an adult ostrich, the openings of the tympanic recesses divide the middle ear into several compartments. The window for the anterior tympanic recess is large and includes the opening of the pharyngotympanic tube. A horizontal bony septum separates the anterior tympanic recess and the interaural pathway from the lateral opening of the pharyngotympanic tube. The actual tympanic

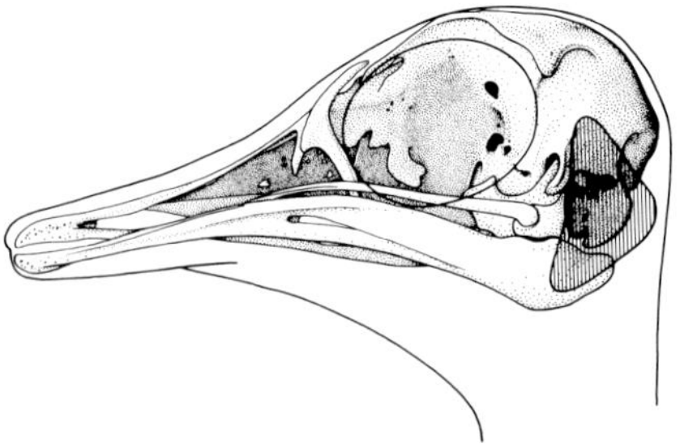

Fig. 27. Ostrich, adult. Lateral view of the skull

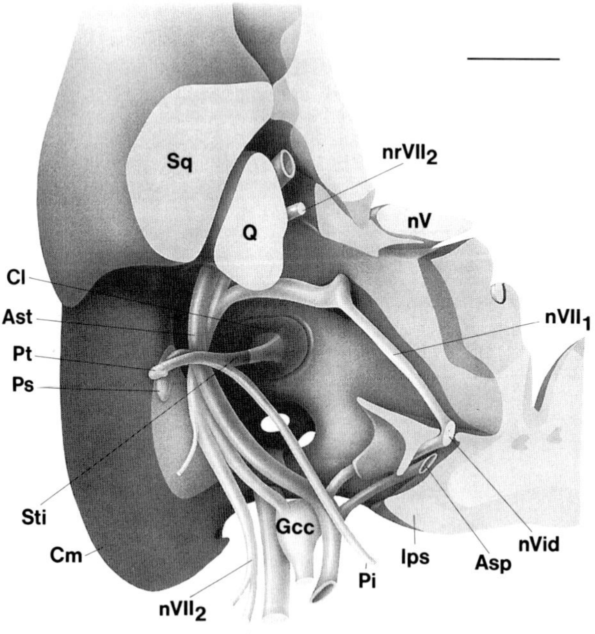

Fig. 28. Ostrich. Plate reconstruction of the right middle ear and basicranium of a hatchling. Section thickness for reconstruction, 80 μm. *Right*, rostral; *left*, occipital; *up*, right; fronto-lateral view; scale bar = 1 mm

cavity is rather small. It is separated from the anterior tympanic recess by a perpendicular bony septum. The subdivisions of the tympanic cavity in the ostrich occur through ossification of connective tissue components; they consist of trabecular bone and are irregularly shaped. Consequently, a high individual variablity of the septa can be observed.

Barred Button Quail. The osteology of the tympanic region of the neognathous button quail differs in many aspects from that described for the paleognathous birds (Fig. 29). As mentioned above, the tympanic cavity faces the lateral side, with the tympanic membrane oriented more or less perpendicularly.

The large tympanic wing of the exoccipital bone forms the occipital wall of the tympanic cavity. Its ventral edge bends rostrally and covers the tympanic membrane with a lateral bony shield. This ventral part also reaches far rostrally, closing the ventral side of the tympanic cavity. In the hatchling, the lateral and ventral parts of the metotic cartilage are still cartilaginous. There is a gradual transition from cartilage to dense connective tissue, which constitutes the ventral wall of the tympanic cavity. This layer of dense connective tissue is connected rostrally to the basitemporal plate. All layers of connective tissue will ossify by the time the bird reaches adulthood. The ossification of connective tissue includes the connective tissue surrounding the internal carotid and the stapedial artery. Therefore, windows for arteries and nerves are found in the ventral bony closure of the tympanic cavity of an adult turnix. In

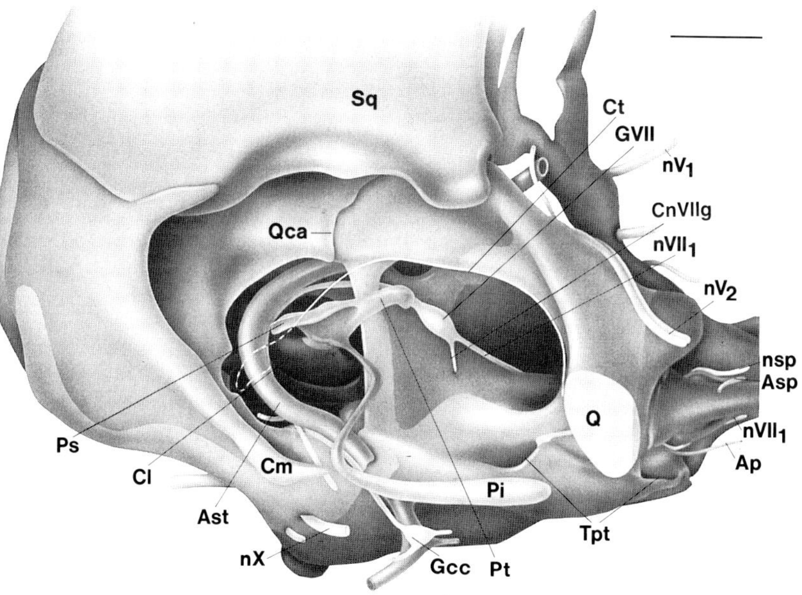

Fig. 29. Barred button quail. Plate reconstruction of the right middle ear and basicranium of a hatchling. Section thickness for reconstruction, 40 μm. *Right*, rostral; *left*, occipital; *up*, right; lateral view; scale bar = 1 mm

turnix, the ventral closure also includes elements of the extracolumella. In the hatchling, the "stylohyal cartilage" is embedded into the plate of dense connective tissue. In the adult, it is ossified and integrated into the ventral closure of the tympanic region. The ossified stylohyal of button quails has previously been described as " lateral basitemporal process" (Bock and McEvey 1969).

Rostrally, the basitemporal plate contributes to the anterior wall of the tympanic cavity. The basitemporal plate is highly pneumatized in turnix, and the entire anterior part of the tympanic cavity is occupied by the large entrances into the anterior tympanic recess and the pharyngotympanic tube. The pharyngotympanic tube opens into a bony tunnel, which is covered ventrally by a wing of the basiparasphenoidal bone.

3.5 Tympanic Membrane (Membrana tympanica): Position, Orientation, Fixation, and Histology

The tympanic membrane is pushed out into the external auditory meatus by the columella. In all paleognathous species, the tympanic membrane is oriented horizontally, with the tip pointing down into the external auditory meatus (Figs. 6–10; Fig. 12D,E).

American Rhea. In the American rhea, the horizontal tympanic membrane is inclined 15° laterally; it is somewhat steeper in the hatchling. It is oval-shaped, the short axis being about two-thirds of the long axis. The long axis of the tympanic membrane is slightly inclined with reference to the mediosagittal axis of the head. The ascending ligament (see following) is visible in macroscopic dissections, leading from the mediocaudal corner of the tympanic membrane to the tip of the tympanic process. The distance between the left and right points of the tympanic membrane was measured as 28.4 mm. Fig. 6A is a ventral view of a macroscopic dissection of the external auditory meatus and the tympanic membranes. All topographic measurements are given in Fig. 6B. The area of the tympanic membrane, when projected onto a plane, was 128 mm^2 in the left and 123.9 mm^2 in the right ear (Fig. 30).

The fixation of the tympanic membrane involves several bones. Caudally, it is fixed to the tympanic wing of the exoccipital bone (cartilago metotica in the hatchling). Laterally, the tympanic membrane is fixed to the lateral edge of the squamosal bone, beneath the pneumatic opening for the superior tympanic recess. The fixation consists of dense connective tissue forming a small ridge at the margin of the tympanic membrane. Rostrally, the tympanic membrane bridges the gap between the squamosal bone and the otic process of the quadrate. At this position it is fixed to the ventral edge of the otic process. Further rostrally, it bridges the gap between the otic process and the lateral wing of the basitemporal plate (lateral wing of the basiparasphenoid in the hatchling). In this anterolateral position, the tympanic membrane is not directly fixed to the wall of the tympanic cavity but rather to dense connective tissue, which isolates it from the kinesis of the otic process. Rostrally, the tympanic membrane is fixed to the ventral edge of the lateral wing of the basitemporal plate (Fig. 34). The lateral opening of the pharyngotympanic tube is included in this fixation. Medially, the tympanic membrane has no direct fixation to the basitemporal plate. In this area the stapedial artery enters the tympanic cavity and the hyomandibular branch of the facial nerve leaves it (see left side of Fig. 31). The stapedial muscle inserts into the tympanic membrane on its medial margin. In the American rhea, the infracolumellar process of the extracolumella ends close to the medial margin of the tympanic membrane. The stylohyal cartilage is embedded in connective tissue of the ventral closure of the tympanic cavity; however, it is located outside of the auditory cavity.

Crested Tinamou. The position of the tympanic membrane resembles that of the American rhea, but the angle to the horizontal plane is 35° to 40°. As in the American rhea, the tympanic membrane bulges out into the external auditory meatus. All measurements are listed in Fig. 7A,B. The area of the tympanic membrane, when projected onto a plane, was 25.4 mm^2 for the left and 26.6 mm^2 for the right ear (Fig. 30).

The fixation of the tympanic membrane is similar to that of the American rhea, although it is tight at the tympanic wing of the exoccipital and the squamosal bone and rather loose where the tympanic membrane bridges the gaps to the otic process of the quadrate.

The connection between the tympanic membrane and the infracolumellar processes of the extracolumella is peculiar in the tinamous. In the crested tinamous (as confirmed in *Tinamou tao*), the distal end of the processus infracolumellaris branches off into two bars, one directed rostrally, the other caudally. Both branches insert into the tympanic membrane for most of their length. The tympanic process is fixed to the tympanic membrane and inserts into the ascending ligament. The superior process of the extracolumella inserts into the tympanic membrane near its posterior margin.

Emu. The tympanic membrane of the emu inclines at a smooth angle of ca. 40° to the horizontal plane. The maximum length of the left eardrum was measured as 14.4 mm, its maximum width as 10.5 mm. A distance of 32.8 mm was measured between the peak points of the left and right tympanic membrane. However, this value is based on calculations from CT scans and thus differs methodically from the direct measurements performed in the other species. An area of 113.6 mm^2 was determined for the left eardrum (Fig. 30).

The fixation of the tympanic membrane of the emu resembles that of the American rhea and involves the same skeletal elements (Fig. 11). The processes of the extracolumella are fixed into the tympanic membrane as described for the American rhea. However, the distal part of the infracolumellar process pierces the tympanic membrane, and its end lies embedded in loose connective tissue outside the tympanic cavity. No isolated stylohyal cartilage has been found.

Cassowary. The tympanic membrane of the cassowary differs from that of the previously described species by its oval-to-round shape. It faces ventrally and inclines laterally at an angle of 35° to the horizontal plane. Linear measurements are given in Fig. 8B. The areas of the tympanic membranes measured 125.9 mm^2 and 126.8 mm^2, respectively (Fig. 30).

Dromaius novae-hollandiae

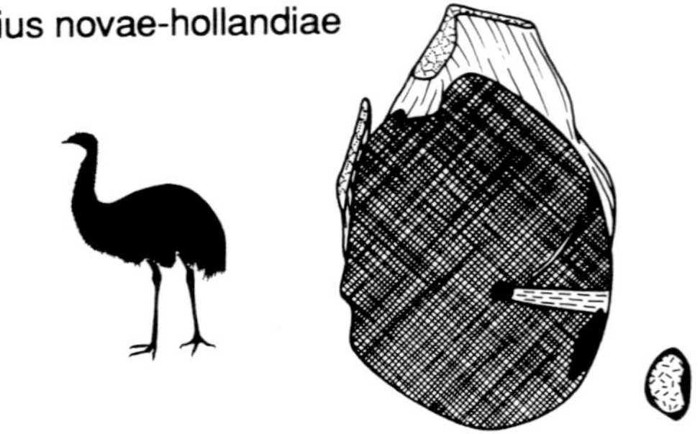

	left	right
Area of tympanic membrane	113.6 mm^2	
Area of clipeolus	3.6 mm^2	
Area ratio of tympanic membrane and clipeolus	31	

Apteryx australis

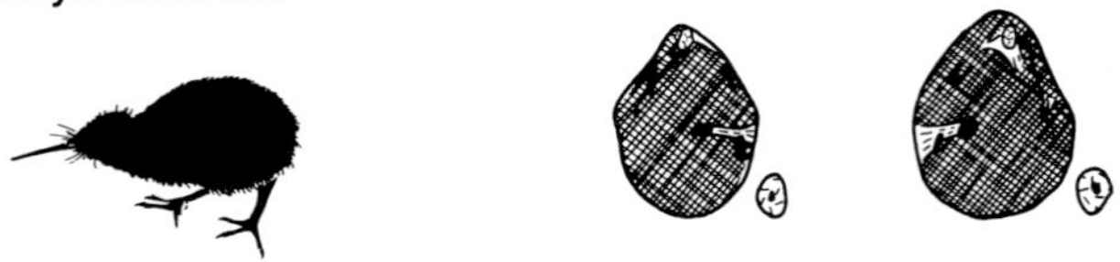

	left	right
Area of tympanic membrane	33.1 mm^2	27.1 mm^2
Area of clipeolus	1.5 mm^2	1.4 mm^2
Area ratio of tympanic membrane and clipeolus	23	18

Casuarius bennetti

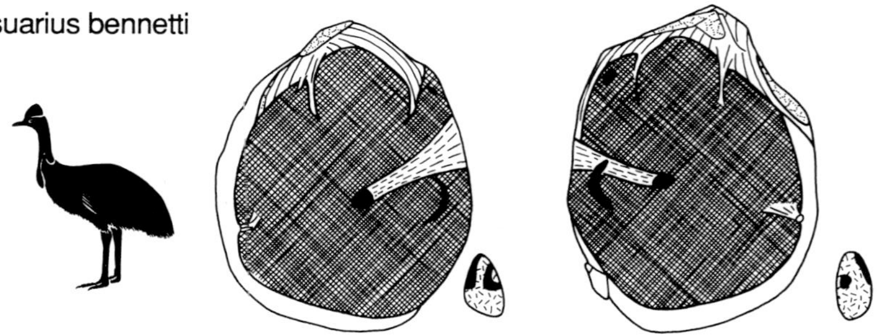

	left	right
Area of tympanic membrane	125.9 mm^2	126.8 mm^2
Area of clipeolus	4.5 mm^2	4.7 mm^2
Area ratio of tympanic membrane and clipeolus	28	27

Turnix suscitator

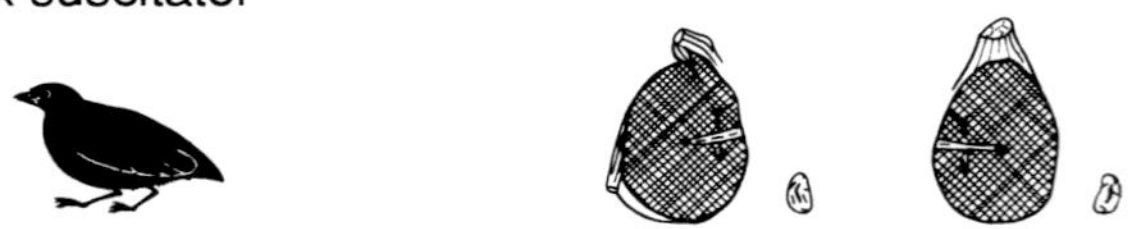

	left	right
Area of tympanic membrane	13.7 mm^2	13.0 mm^2
Area of clipeolus	0.6 mm^2	0.6 mm^2
Area ratio of tympanic membrane and clipeolus	22	22

Struthio camelus

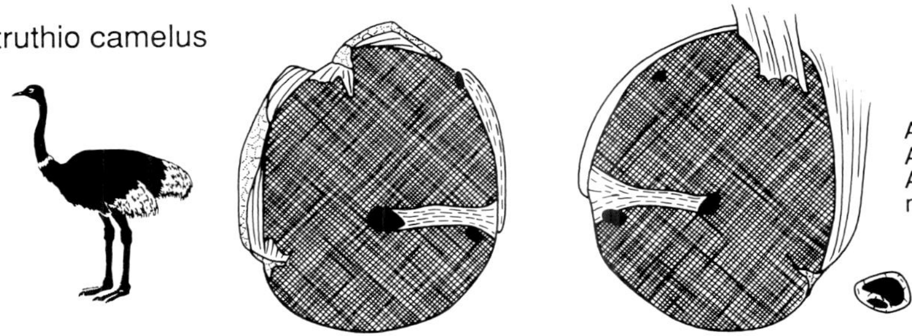

	left	right
Area of tympanic membrane	134.5 mm²	146.0 mm²
Area of clipeolus		3.9 mm²
Area ratio of tympanic membrane and clipeolus		37

Rhea americana

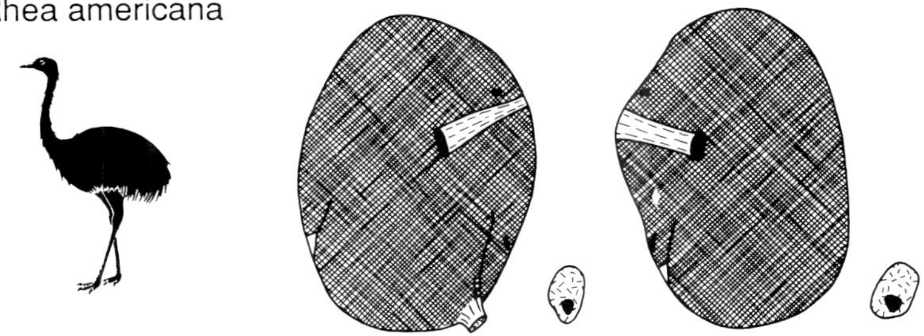

	left	right
Area of tympanic membrane	128.0 mm²	123.9 mm²
Area of clipeolus	3.5 mm²	4.2 mm²
Area ratio of tympanic membrane and clipeolus	37	30

Eudromia elegans

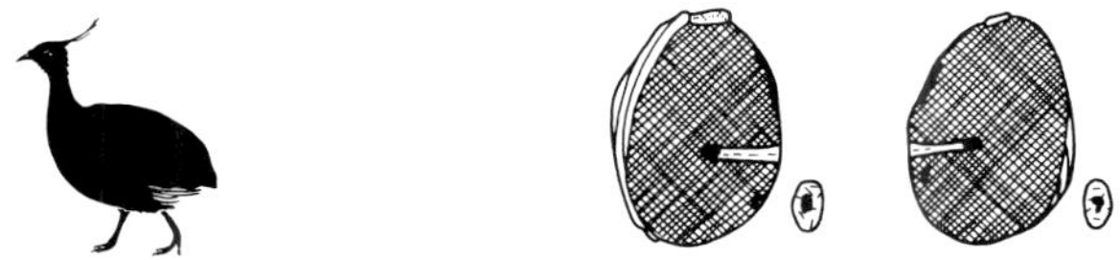

	left	right
Area of tympanic membrane	25.4 mm²	26.6 mm²
Area of clipeolus	1.0 mm²	1.0 mm²
Area ratio of tympanic membrane and clipeolus	25	26

Fig. 30. Morphometry of the tympanic membrane and the clipeolus. Planar projections of the left and right tympanic membrane are given as measured. Position of the ascending ligament, insertion of the stapedial muscle, and insertion point of the three branches of the extracolumella are shown (camera lucida drawings from preserved specimens). Beside each tympanic membrane the area of the clipeolus of its columella is depicted. Area measurements of the tympanic membrane, the clipeolus, and the area ratio of the tympanic membrane and clipeolus are given in tables on the *right*

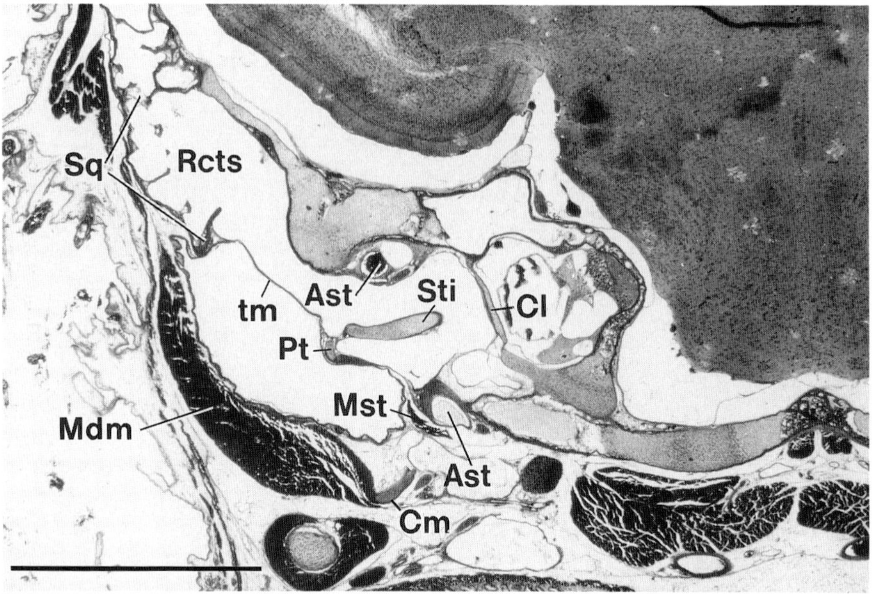

Fig. 31. Barred button quail. Cross section through the head of a hatchling, microphotograph of the otic region. Paraffin, serial section #36/20, 15 μm, azan, scale bar = 1 mm

The fixation of the tympanic membrane resembles that described for the American rhea. The extracolumella connects to the tympanic membrane with the tips of its three processes. The knob-shaped tip of the tympanic process is tightly inserted into the ascending ligament and also connects to some collagenous fibers originating from the stapedial muscle. The tympanic end of the supracolumellar process is a triangle-shaped cartilaginous plate that inserts into the eardrum with its distal edge. The short infracolumellar process inserts into the ventromedial margin of the tympanic membrane. Its distal end lies outside the tympanic cavity, close to the stapedial artery (see following).

Kiwi. Both tympanic membranes of the kiwi are rather irregularly shaped and more or less round. They are inclined at an angle of 35° to the horizontal plane. The measurements were 8.0 mm × 5.6 mm for the left and 7.6 mm × 5.6 mm for the right side. The pointed tips of the tympanic membranes were 20.0 mm apart (Fig. 9). Eardrum areas were 33.1 mm^2 and 27.1 mm^2, respectively (see Fig. 30).

The fixation of the tympanic membrane to the margin of the tympanic cavity in the kiwi resembles that of the rhea. Minor differences can be found in the fixation to the processes of the extracolumella. The tympanic process of the extracolumella is fixed to the tympanic membrane and inserts into the distal end ot the ascending ligament. The short supracolumellar process also inserts into the ascending ligament, only a short distance proximal to the insertion of the tympanic process. The infracolumellar process is attached close to the margin of the tympanic membrane. The tip of the infracolumellar process pierces the membrane and ends in loose connective tissue surrounding the tympanic cavity.

Ostrich. The position and orientation of the tympanic membrane of the ostrich resemble those of the other paleognathous species. It has a highly oval form and inclines at an angle of 30°–40° to the lateral side. The longitudinal axes of the tympanic membranes are inclined at an angle of 31° to the sagittal midline of the head. The linear measurements are given in Fig. 10. The areas for the left and right sides were 134.5 mm^2 and 146.0 mm^2 (Fig. 30).

The tympanic membrane of the ostrich is peculiar in that it forms a fold of connective tissue. This fold originates at the rostral margin of the tympanic membrane near the opening of the pharyngotympanic tube. There it forms a small pocket. Further caudally, it becomes enlarged, emerging from the dorsolateral part of the tympanic membrane, close to the otic process of the quadrate. It is attached to the ventromedial corner of the tympanic cavity. Its lateral part connects to the otic process of the quadrate bone and fuses with Platner's ligament, which can be distinguished histologically since it contains elastic fibers.

Barred Button Quail. The eardrum of the neognathous barred button quail differs in some aspects from those of paleognathous species. The tympanic membrane is oriented to the lateral side of the head, approximately perpendicular to the horizontal plane. Although the button quail is a very small bird, the oval tympanic membrane measures between 4.6 mm in maximum length and 3.6 mm for the maximum width in the left and right eardrums. Tympanic membrane areas were determined to be 13.7 mm^2 in the left and 13.0 mm^2 in the right ear (Fig. 30).

The fixation of the tympanic membrane differs from that of the paleognaths. At its caudal and ventral margin, the tympanic membrane is fixed to the metotic cartilage (Fig. 31). The metotic cartilage extends far to the lateral and ventral side of the skull, which affects the fixation of the ear drum. The metotic cartilage replaces the fixation to the basitemporal plate, which has been described in the paleognathous species. Dorsally, the eardrum connects to the lateral and ventral edge of the squamosal bone. In the hatchling, the dorsorostral fixation consists of dense connective tissue stretching between the tympanic membrane and the otic process of the quadrate. Rostroventrally, the tympanic membrane is not directly fixed to the skull but ends in dense connective tissue. In the hatchling, dense connective tissue closes the middle ear cavity ventrally. All connective tissue ossifies later in ontogeny to become an osseous wall of the middle ear and a bony fixation of the tympanic membrane. The stapedial artery, the hyomandibular branch of the facial nerve, the glossopharyngeal nerve, and parasympathetic nerves pierce this bony plate (Fig. 29).

The distal ends of the processes of the extracolumella are tightly fixed to the eardrum. The fixations of the tympanic process and of both prongs of the supracolumellar process reveal no substantial differences from the paleognathous species, except for the fact that both supracolumellar prongs are present in the hatchling. However, the fixation of the infracolumellar process

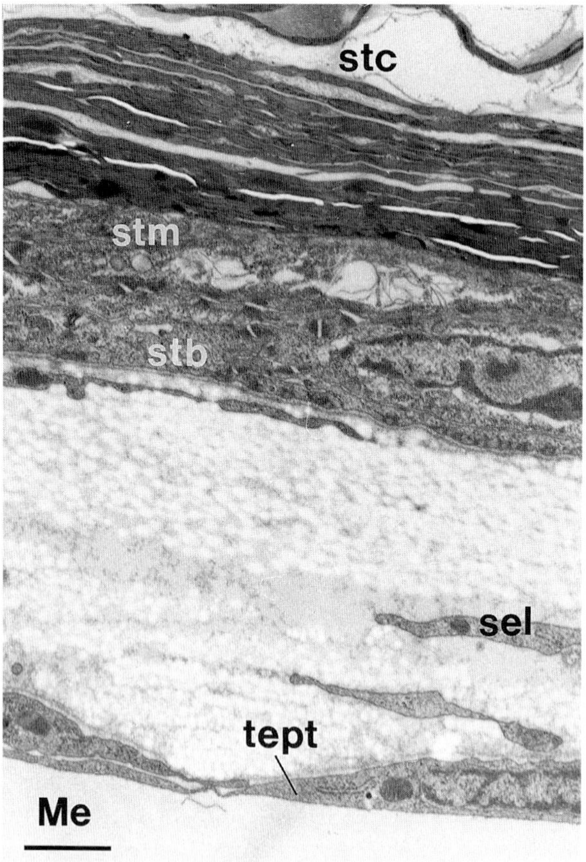

A

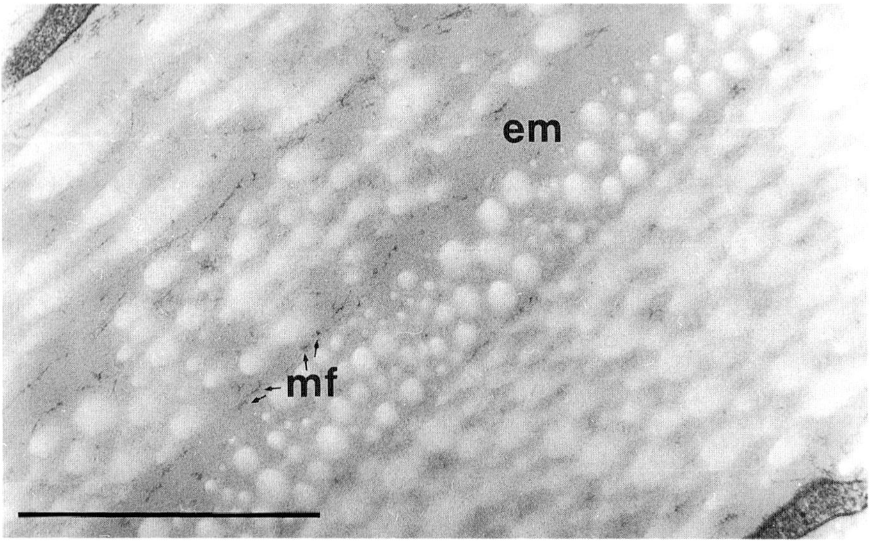

B

Fig. 32A,B. Histology of the tympanic membrane of domestic fowl. **A** Low magnification shows the three layers: epidermis, connective tissue layer, and internal epithelium; scale bar = 1 μm. **B** The connective tissue layer reveals a thick, amorphous extracellular matrix and microfibrillar components; scale bar = 1 μm

to the tympanic membrane differs in many aspects. The distal part of the infracolumellar process is embedded in dense connective tissue of the ventral wall of the tympanic cavity. During posthatching ontogeny, both elements ossify (for details see section 3.10).

Histology. The histology of the tympanic membrane is identical in all species studied. It has three layers: (1) the external epidermal layer, (2) a middle layer of connective tissue, with few capillaries and nerves, and (3) the internal (tympanic) single-layered epithelium. Its ultrastructure has been documented in domestic fowl (Fig. 32A,B).

1. The epidermal layer is a cornified epithelium similar to that of the body surface, but thinner than at other locations. The flat, elongated cells of the stratum basale are arranged on a basal lamina. A two-cell-thick intermediate layer (stratum transitivum) follows towards the external side, with a gradual transition to the ceratin layer (stratum corneum). Amorphous deposits of wax are found between the cells of the keratin layer.
2. The middle, subepidermal layer of the tympanic membrane consists of connective tissue. Ultrastructural investigations of this layer reveal a high proportion of extracellular matrix and microfibrillar components resembling elastin (Fig. 32B). Thus, the ultrastructural findings imply a high compliance of the tympanic membrane. The connective tissue layer also contains melanocytes and capillaries. The ascending ligament (see following) is embedded in this layer. The thickness of the middle layer changes from the margins, where it is thick, towards the central parts, where it becomes thinner. All processes of the extracolumella, the stapedial muscle, and the ascending ligament insert into the middle layer of the tympanic membrane. Preliminary investigations of the elastic fiber distribution of the tympanic membrane could not detect a regular pattern of arrangement.
3. The inner layer of the tympanic membrane is a thin, single-layered epithelium of flattened cells (Fig. 32A,B) on a basal lamina. Its cells have finger-like processes touching fibroblasts of the middle layer. The inner epithelium covers not only the tympanic surface of the eardrum but also continues over all middle ear structures including the stapes and the extracolumella. Thus, a thin, two-layered, epithelial fold stretches between the processes of the extracolumella and the tympanic membrane. It continues into the epithelium of the pharyngotympanic tube and the pharynx.

3.6 Tympanic Recesses and Pneumatization of the Neurocranium

The middle ear of birds has three tympanic recesses: recessus tympanicus anterior, rec. tymp. superior, and rec. tymp. posterior. All recesses enter the bony skull, where they form large pneumatic spaces. The terms "recess" and "pneumatic space" refer to the same anatomical elements although "pneumatic space" is less accurate and may be used in other contexts as well. The nomenclature is based on the position of their openings in the tympanic cavity. The superior tympanic recess (rec. tympanicus superior) intrudes into the squamosal bone; the posterior tympanic recess (rec. tympanicus posterior) pneumatizes the exoccipital, supraoccipital, and prootic bones; and the anterior tympanic recess (rec. tympanicus anterior) pneumatizes the anterior parts of the skull and connects the contralateral middle ear cavities. The air volumes of the middle ear and the adjacent pneumatic spaces contribute stiffness reactance and resonance properties to the transfer function of the middle ear. Thus, the structural and morphometrical investigation of the tympanic recesses and penumatic spaces of the avian skull may provide data for functional morphological models. However, until recently the analysis of nonstructures such as air-filled spaces presented considerable technical problems. Therefore, most descriptions in the literature refer to opening foramina of the recesses but not to their form and position in the skull. Now, noninvasive X-ray computed tomography produces detailed images of air spaces which can be processed by quantitative image analysis. Computer aided three-dimensional reconstructions are a tool for precise anatomical analysis of these cavities.

American Rhea. The superior tympanic recess originates from the lateral and dorsal part of the tympanic cavity. Two large pneumatic windows open on the ventral side of the squamosal, from where the recess intrudes and totally pneumatizes this bone. Furthermore, in an adult rhea, most of the lateral bony wall of the skull is pneumatized (Fig. 12B–E). Several smaller diverticles emerge from the main pneumatic volume, intrude into all other parts of the skull, and fuse with other recesses (Fig. 33). The superior tympanic recess is filled with a trabecular network, which is not shown in Fig. 33. The recess and the pneumatization of squamosal are already present in the hatchling (Fig. 34). However, in a hatchling only one pneumatic window opens into the squamosal.

The posterior tympanic recess has several openings in the rear part of the tympanic cavity. A larger window is found in the metotic cavity, between the otic capsule and the lateral wing of the exoccipital (Fig. 20). Additionally, several smaller windows are found in a shallow oval depression on the lateral wing of the exoccipital behind the cranial articulation facet for the quadrate bone. In the adult rhea, the superior and the posterior tympanic recesses are fused and it is difficult to separate them. Only a small bony ridge separates the openings of the posterior tympanic recess from those of the superior tympanic recess, and indicates the border between exoccipital and squamosal bones. Pneumatizations of the posterior tympanic recess extend throughout most of the occipital region, consisting of the exoccipital, the supraoccipital, and the basioccipital bone. The contralateral recesses are not fused (Fig. 33). In the hatchling, only one window of the posterior tympanic recess is found in the metotic cavity. The metotic cartilage is still solid and has no pneumatic perforations.

The anterior tympanic recess emerges from the rostral part of the tympanic cavity. Its funnel-shaped opening occupies the anterior half of the tympanic cavity (Fig. 12B,C; Figs. 19, 33, 34). It pneumatizes the basitemporal plate. The contralateral recesses fuse in the midline of the skull and connect the left and right middle ear (Fig. 12B,C; Figs. 33, 34). This connection, the interaural pathway, is a shallow tunnel between both tympanic cavities. From the interaural pathway, further pneumatizations intrude rostrally into the rostrum of the parasphenoidal bone and caudally into the basitemporal plate.

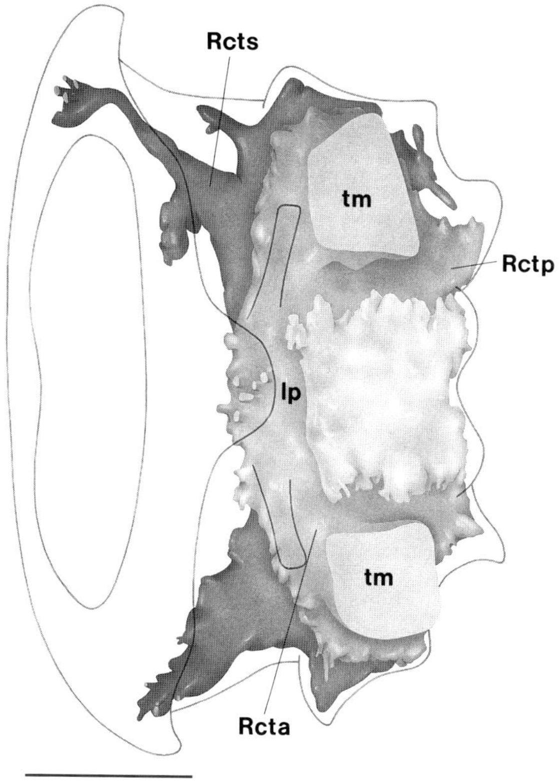

Fig. 33. American rhea, adult. Reconstruction of the tympanic spaces, the interaural pathway, and adjacent pneumatic spaces of the otic region of the skull. Computer-aided reconstruction of 70 equidistant CT scans of 1-mm thickness. Outline of the skull is represented by a line; all air spaces are presented in gray. Scale bar = 10 mm; ventral view

The volume of the tympanic cavities and the adjacent pneumatic spaces has been measured using X-ray computed tomography and computer-aided morphometry. In the American rhea, the air volume of the two tympanic cavities was measured as 19 453 mm^3. The tympanic cavities and pneumatic spaces added up to a total air volume of 46 634 mm^3.

Crested Tinamou. The same pattern of tympanic recesses and pneumatic spaces is present in the crested tinamou. The superior tympanic recess enters the squamosal bone through a large window (Fig. 12). However, the superior tympanic recess is rather small in comparison to that of the American rhea. In adult birds it is completely fused with the posterior tympanic recess (Fig. 35), and can be recognized only by its foramen in the squamosal bone.

The topography of the openings of the posterior tympanic recess is similar to that of the American rhea. A large window in the rear part of the metotic cavity is already present in the hatchling. Additional windows develop after ossification of the metotic cartilage in the lateral wing of the exoccipital bone. The pneumatic spaces emerging from the posterior tympanic recess pneumatize the exoccipital, the basioccipital, and the supraoccipital bone. The dorsal parts

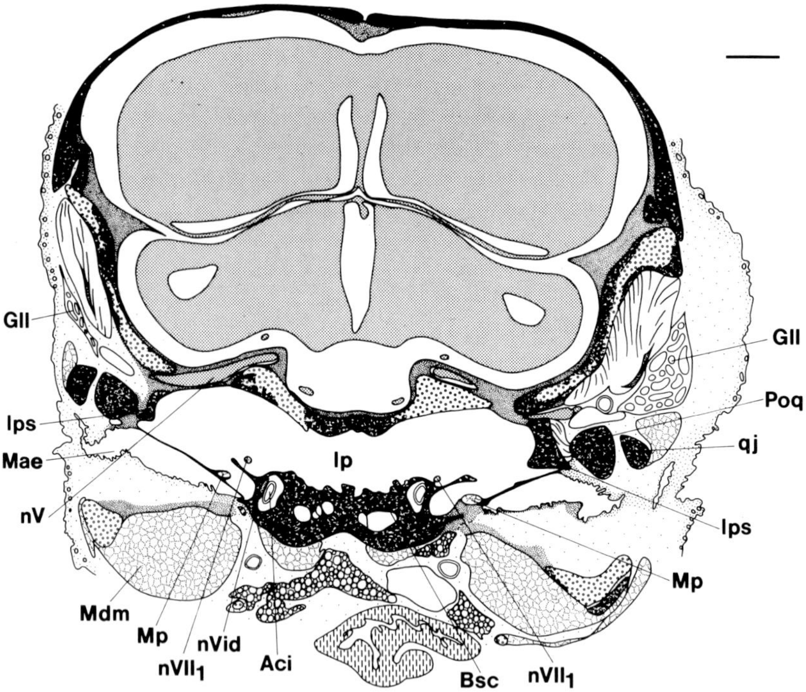

Fig. 34. American rhea, hatchling. Cross section through the head. Camera lucida drawing of the region of the recessus tympanicus anterior (interaural pathway). Celloidin, serial section #330/1, 40 μm, azan, scale bar = 2 mm

of the contralateral posterior tympanic recesses are fused in the dorsal midline above the occipital foramen in the supraoccipital bone (Fig. 35).

The anterior tympanic recess is the largest recess and emerges from the frontal part of the tympanic cavity. The entrance to the recess is funnel-shaped and opens into the basitemporal plate. The basitemporal plate is completely pneumatized by the anterior process. The anterior tympanic recesses of both sides fuse in the midline of the basitemporal plate, forming the interaural pathway (Fig. 13B–E; Figs. 35, 36), which is large and wide, with an extension entering the parasphenoidal rostrum.

The volume of the tympanic cavities and the interaural pathway was calculated as 3057 mm^3, 2421 mm^3 was determined for the remaining pneumatic spaces. Thus, the skull enclosed a total air volume of 5478 mm^3.

Emu. The same pattern of three tympanic recesses is present in the emu. The position and development of pneumatic windows are no different from those of the rhea. However, in an adult emu, all recesses and their pneumatizations are fused into one large and extended pneumatic space in the lateral wall of the skull and the basicranium (Figs. 14, 37). Thus, the recesses cannot be separated topographically. However, a broad interaural pathway is established connecting both middle ear cavities through the basitemporal plate. As in the tinamou or rhea, no supracerebral connection is present.

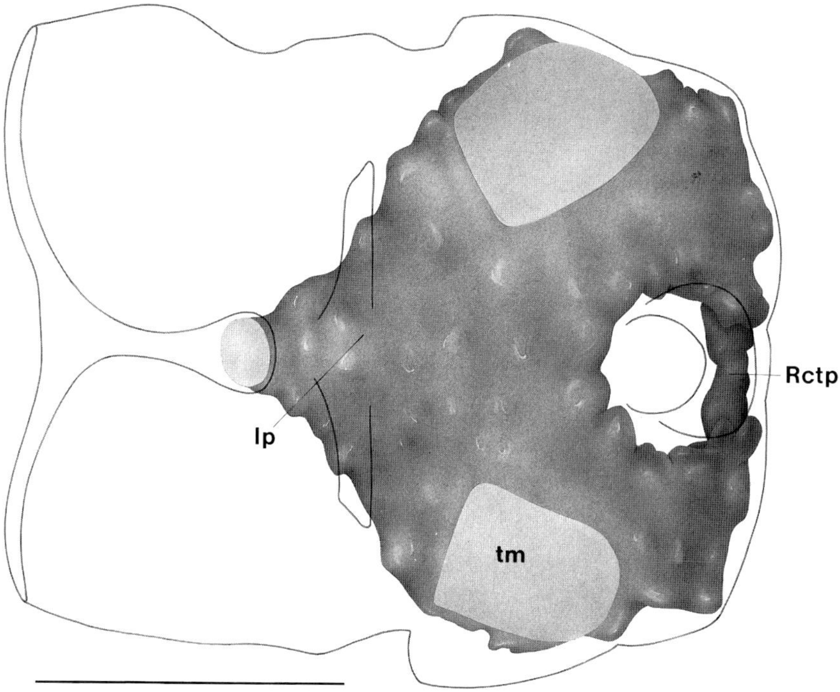

Fig. 35. Crested tinamou, adult. Reconstruction of the tympanic spaces, the interaural pathway, and adjacent pneumatic spaces of the otic region of the skull. Computer-aided reconstruction of 25 equidistant CT scans of 1-mm thickness. Outline of the skull is represented by a line; all air spaces are presented in gray. Scale bar = 10 mm; ventral view

Because of the wide pneumatic windows, it was not possible to differentiate between tympanic and recess volume. The total air volume of the tympanic cavity and adjacent pneumatic spaces was determined as 40 426 mm^3.

Cassowary. The large pneumatic spaces in the casque of the dwarf cassowary modify the pattern of pneumatization to a rather confusing system of air tubes and spaces. Despite this modification, the general arrangement of three tympanic recesses and an interaural pathway in the basitemporal plate can be recognized. In the adult dwarf cassowary, the openings of the superior tympanic recess and the posterior recesses emerge together from a shallow bony pit extending from the ventral side of the squamosal to the lateral wing of the exoccipital. Many small windows open from this pit (Fig. 15B–E). However, it is not possible to differentiate between openings of the superior and the posterior tympanic recess in adult specimens. In the hatchling southern cassowary, both recesses are still separate, with the opening for the superior recess in the squamosal bone and that for the posterior recess in the metotic cavity. The superior tympanic recess pneumatizes not only the squamosal bone but also the parietal and frontal bone and dorsally connects to the large pneumatic space of the casque. Several smaller air spaces emerge from the superior recess and pneumatize the lateral wall of the skull (Fig. 38A,B).

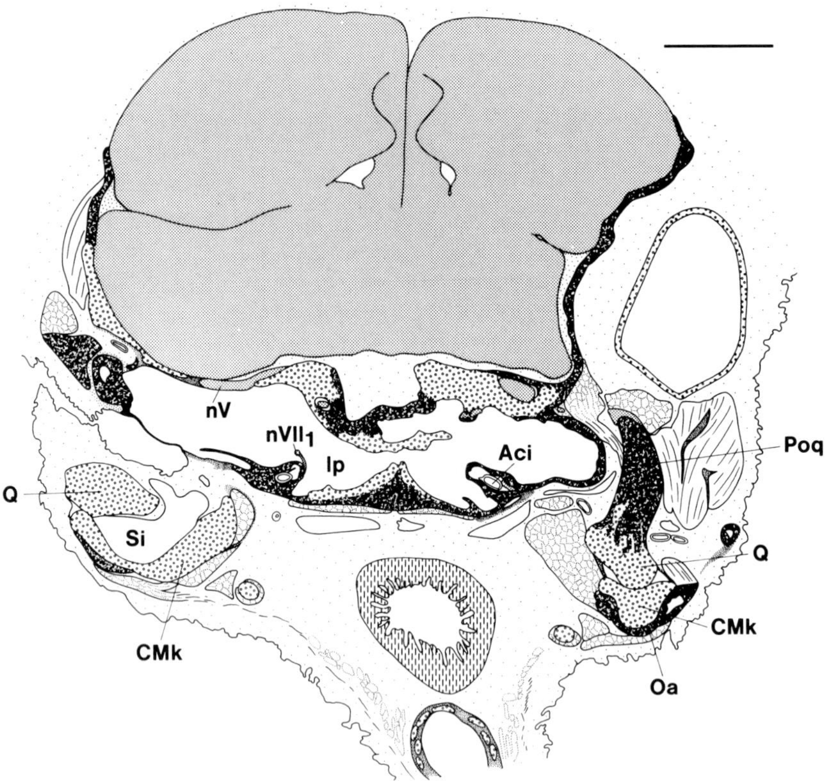

Fig. 36. Crested tinamou, hatchling, cross section through the head. Camera lucida drawing of the region of the recessus tympanicus anterior (interaural pathway). Paraffin, serial section #Eu3/51/5, 15 μm, azan, scale bar = 2 mm

The posterior tympanic recess of the adult cassowary comprises a large air volume and pneumatizes all occipital bones and the tympanic wing of the exoccipital. Several ascending air tubes are fused to the supracerebral air space in the helmet. Thus, the superior and posterior tympanic recesses as well as the contralateral recesses are inter connected in the pneumatization of the helmet.

The opening of the anterior tympanic recess occupies most of the tympanic cavity. The anterior tympanic recess connects both middle ear cavities. A large diverticle enters the parasphenoidal rostrum. The anterior tympanic recess has no connection with the air space of the helmet.

In the dwarf cassowary, the volume of both middle ears and the interaural pathway comprised 47 620 mm^3. The volume of the pneumatic spaces, including the helmet, was 250 820 mm^3. This is a minimum estimate since only the otic region of the helmet was measured. The total air volume of the skull was greater than 298 067 mm^3.

Kiwi. Although the same pattern of three tympanic recesses can be described, the kiwi shows another extreme. In the adult, the superior recess is very small (Fig. 16E). It can be better recognized in the hatchling, where, as in the other

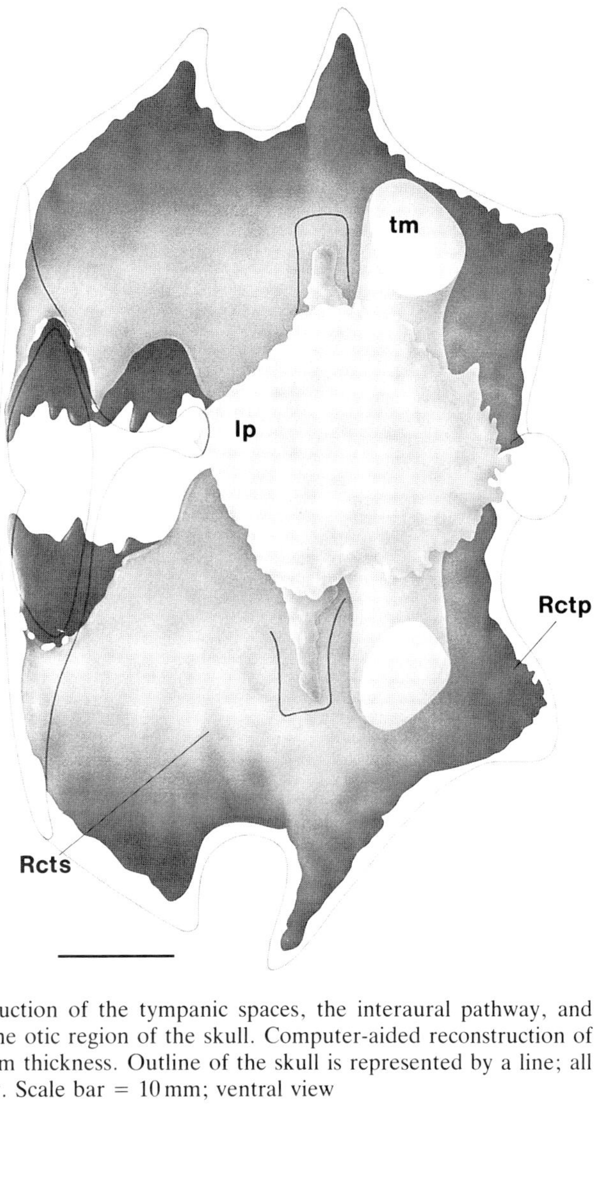

Fig. 37. Emu, adult. Reconstruction of the tympanic spaces, the interaural pathway, and adjacent pneumatic spaces of the otic region of the skull. Computer-aided reconstruction of 79 equidistant CT scans of 1-mm thickness. Outline of the skull is represented by a line; all air spaces are presented in gray. Scale bar = 10 mm; ventral view

species, it opens into the squamosal. At hatching, the posterior tympanic recess has only one window in the metotic cavity. By adulthood, it develops a second metotic and a squamosal window which, however, fuses to the squamosal window of the superior tympanic recess. In adults, it is not possible to separate the windows of both recesses. The superior tympanic recess has a very small volume in the kiwi and is almost impossible to separate from the superior recess since they are broadly fused. Both recesses together form a pneumatic sac reaching into the occipital region of the head. No connection exists between the contralateral recesses. The anterior tympanic recess is well developed in the kiwi and connects left and right tympanic cavities as it does in the other species. No dilatations emerge from the anterior tympanic recess exept

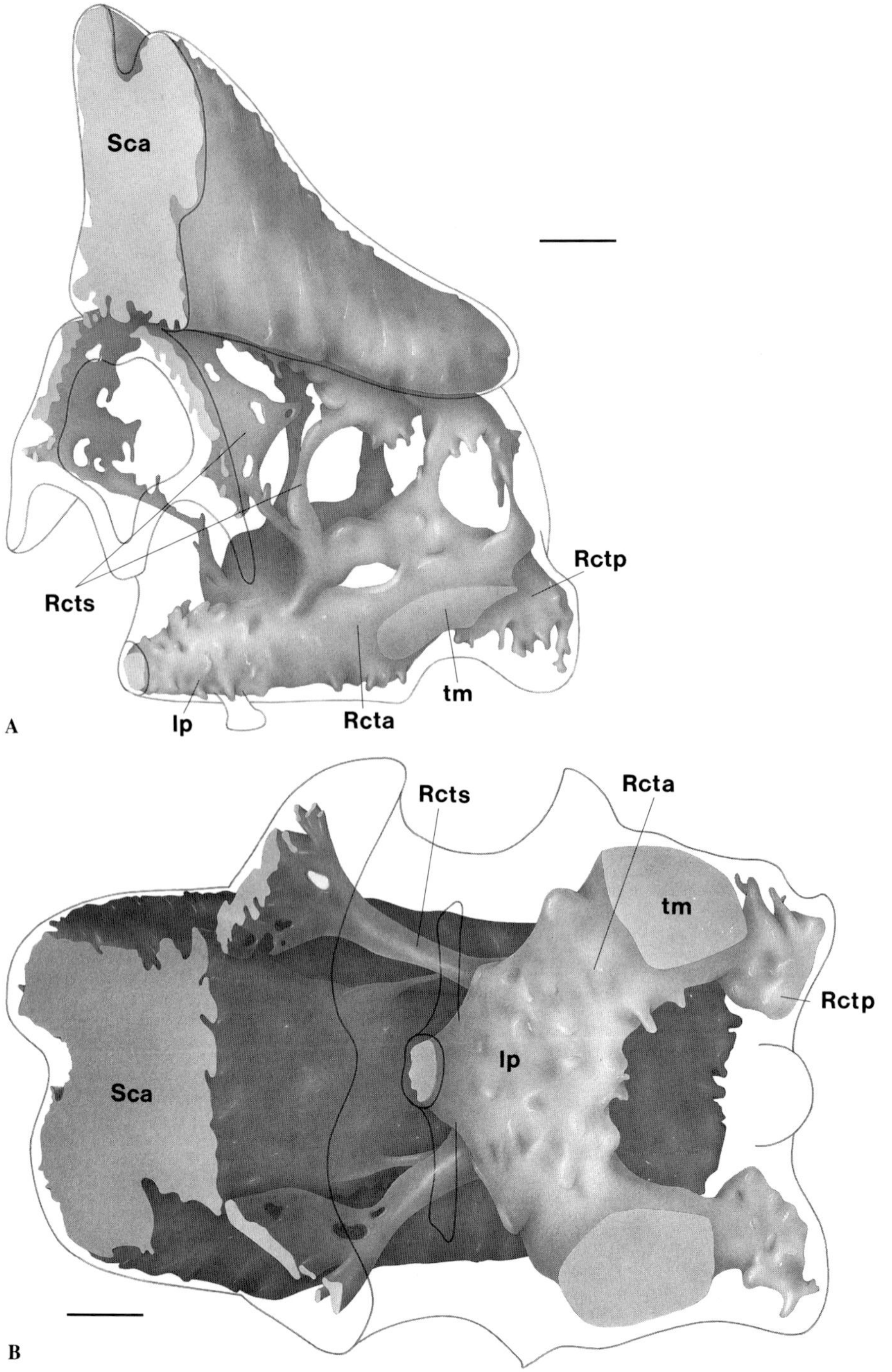

Fig. 38A,B. Dwarf cassowary, adult. Reconstruction of the tympanic spaces, the interaural pathway, and adjacent pneumatic spaces of the otic region of the skull. Computer-aided reconstruction of 39 equidistant CT scans of 1-mm thickness. Outline of the skull is represented by a line; all air spaces are presented in gray. Scale bar = 10 mm. **A** Lateral view. **B** Ventral view

for a small anterior and medial process intruding a short distance into the parasphenoidal rostrum. Thus, all tympanic recesses are present and an interaural pathway is formed (Figs. 16, 39B–E).

It is obvious from the reconstructions and the description just given that pneumatic volumes are small. The volume of middle ear cavities, including the interaural pathway, was determined as 5873 mm^3. The volume of all other pneumatizations of the skull was measured to be 2862 mm^3, resulting in a total air volume of 8735 mm^3.

Ostrich. In the ostrich, the arrangement of the tympanic recesses is similar to that just described. However, a high degree of pneumatization modifies the structure. The superior tympanic recess emerges from the upper part of the tympanic cavity and enters the squamosal bone through an opening on its ventral side. It is a large recess that pneumatizes not only the squamosal but also the parietal and frontal bones (Fig. 40A,B). Dorsal dilatations of the contralateral recesses fuse in the dorsal midline of the skull. Some extensions of the superior tympanic recess reach into the postorbital and zygomatic processes of the skull, thus pneumatizing the squamosal and frontal bone (Fig. 17B–E). The windows to the superior and posterior tympanic recesses of the ostrich are not fused and are easily distinguishable.

The posterior tympanic recess is the largest of the three recesses. It enters the exoccipital bone through one window in the tympanic wing of the

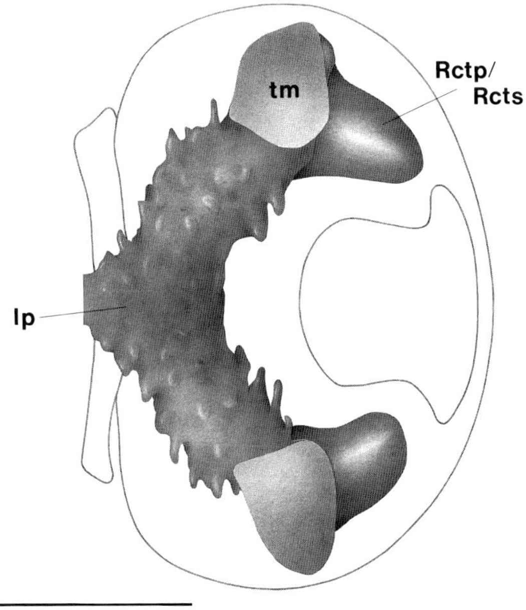

Fig. 39. Kiwi, adult. Reconstruction of the tympanic spaces, the interaural pathway and adjacent pneumatic spaces of the otic region of the skull. Computer-aided reconstruction of 27 equidistant CT scans of 1-mm thickness. Outline of the skull is represented by a line; all air spaces are presented in gray. Scale bar = 10 mm; ventral view

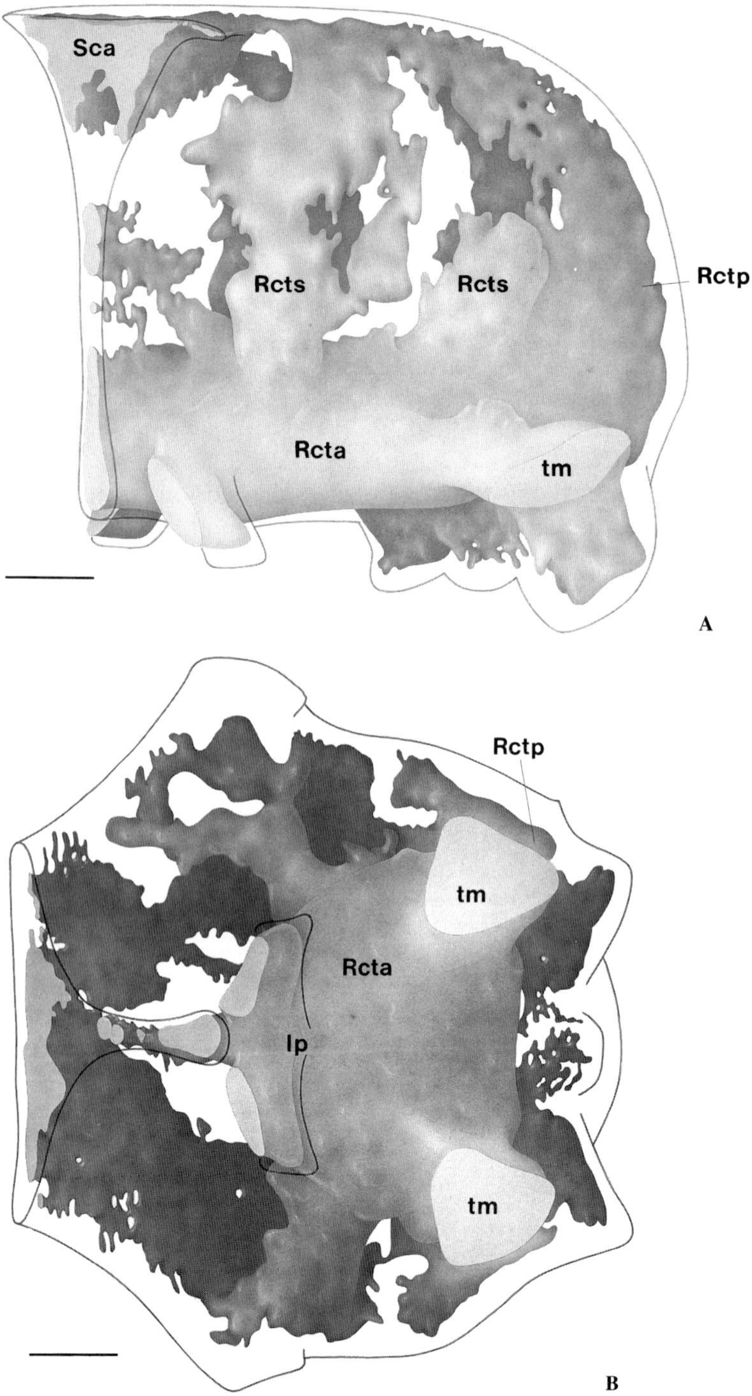

Fig. 40A,B. Ostrich, adult. Reconstruction of the tympanic spaces, the interaural pathway, and adjacent pneumatic spaces of the otic region of the skull. Computer-aided reconstruction of 63 equidistant CT scans of 1-mm thickness. Outline of the skull is represented by a line; all air spaces are presented in gray. Scale bar = 10 mm. **A** Lateral view. **B** Ventral view

exoccipital bone and a second window on the posterior side of the metotic cavity. Both pneumatic windows are clearly separated from the opening of the superior tympanic process. The posterior tympanic recess has three portions: the lateral part pneumatizes the tympanic wing of the exoccipital bone, the upper part enters the parietal region of the skull, and the occipital part pneumatizes the occipital and supraoccipital bones. This occipital part of the posterior tympanic recess connects to the supracerebral air space in the dorsal midline of the skull.

The anatomy of the anterior tympanic recess of the ostrich resembles that of the other species. It connects both middle ear cavities as interaural pathway. Species-specific dilatations emerge from the anterior tympanic recess. Rostrally, the anterior recess pneumatizes the anterior parts of the basitemporal plate, including the basitemporal processes (proc. basipterygoidei) and the parasphenoidal rostrum. Some smaller diverticles enter the broad interorbital septum (Figs. 17B–E, 41). Thus, in the ostrich all tympanic recesses fuse with their contralateral recesses. Functionally, they form not only a ventral but also a dorsal interaural pathway.

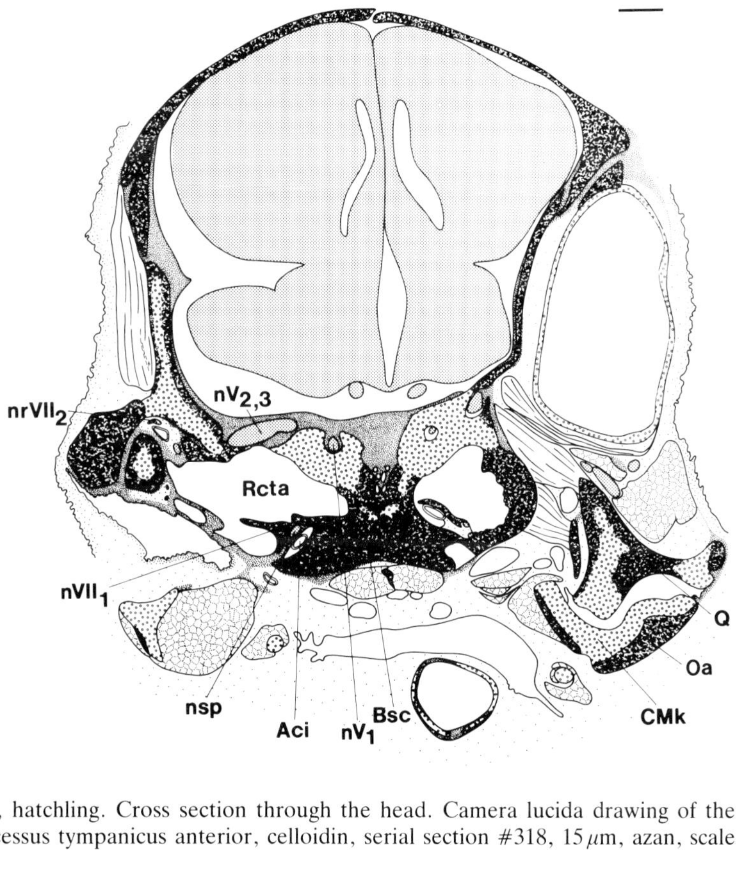

Fig. 41. Ostrich, hatchling. Cross section through the head. Camera lucida drawing of the region of the recessus tympanicus anterior, celloidin, serial section #318, 15 μm, azan, scale bar = 2 mm

The volume of both tympanic cavities and the interaural pathway amounted to 130 297 mm^3. All remaining pneumatic spaces had a volume of 210 152 mm^3, resulting in a total air volume of 340 349 mm^3.

Barred Button Quail. The head of the barred button quail is anatomically characterized by its high degree of pneumatization (Fig. 42). The superior tympanic recess enters the squamosal bone ventrally through one large window. Caudally, the recess fuses with the posterior tympanic recess, and no clear distinction between the two recesses is possible (Fig. 43). The posterior tympanic recess enters the exoccipital bone from the posterior part of the metotic cavity through one wide opening. It pneumatizes all of the occipital region and the otical region. Superior and posterior tympanic recesses of each side are fused and surround the canalicular capsule with pneumatic spaces. The anterior tympanic recess has a large funnel-shaped lateral opening. The contralateral recesses fuse via the wide interaural pathway and pneumatize the parasphenoidal rostrum.

Morphometry of the tympanic cavity and all adjacent pneumatic spaces of the turnix head resulted in a volume of 1560 mm^3.

Comparison and Functional Considerations. In paleognathous birds, the openings of the tympanic recesses show only minor structural variations. The superior tympanic recess has a window into the squamosal, and the posterior tympanic recess has at least two large windows, one in the tympanic wing of the exoccipital bone and a second on the posterior side of the metotic cavity. The windows probably fused with additional smaller windows which may have opened during ontogeny. A much greater variation is present in the arrangement of pneumatic spaces emerging from the tympanic recesses, their volume, and their interconnections. However, it is obvious that the basic pattern of three

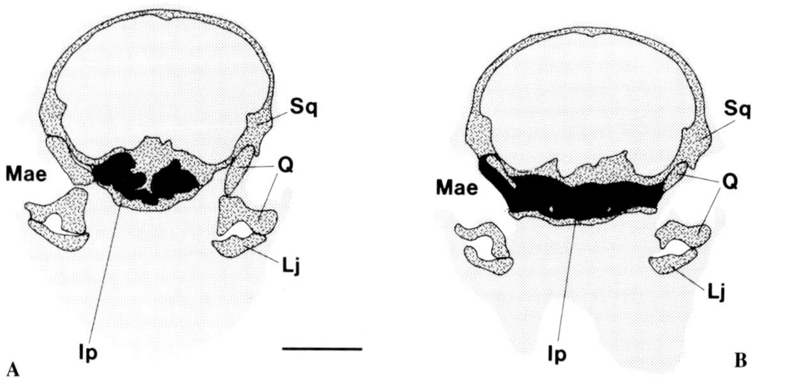

Fig. 42A,B. Barred button quail, adult. Cryostat sections through the otic region of the head. Section thickness 0.2 mm, unstained. Redrawn from microphotographs. **A** and **B** Subsequent levels through the otic region. To demonstrate the interaural pathway, only the anterior tympanic recess is shown here. The superior and posterior tympanic recesses, which pneumatize the whole skull, are not shown. Scale bar = 5 mm

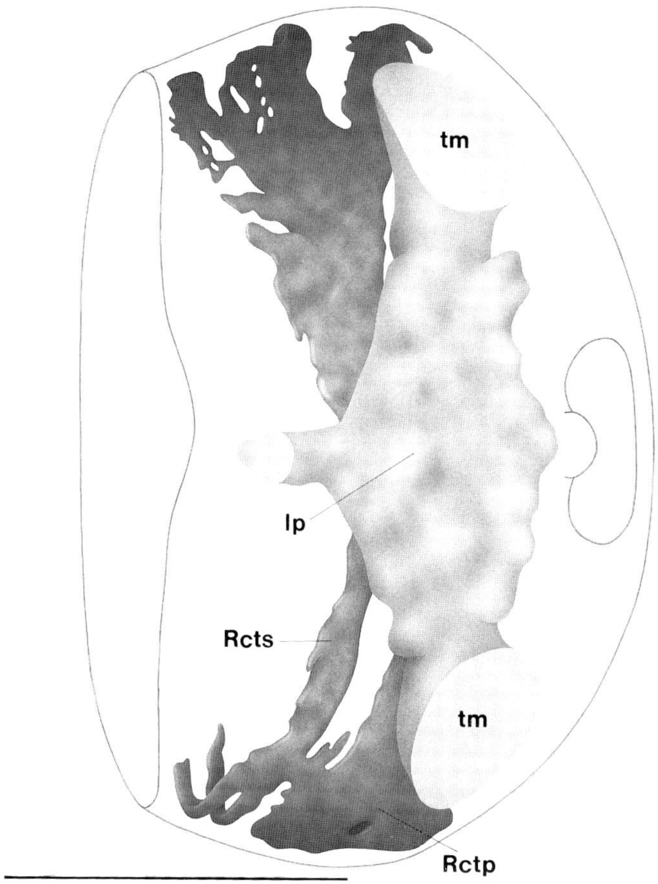

Fig. 43. Barred button quail, adult. Reconstruction of the tympanic spaces, the interaural pathway, and adjacent pneumatic spaces of the otic region of the skull. Computer-aided reconstruction of 50 equidistant cryostat sections of 0.2-mm thickness. Outline of the skull is represented by a line; all air spaces are presented in gray. Scale bar = 10 mm; ventral view

tympanic recesses with an interaural pathway is constant for paleognathous birds as well as for turnix.

The description of three tympanic recesses in birds goes back to Cuvier (1798–1805), Tiedemann (1810), and Gadow (1891). A review of avian ear structures by Freye (1952/1953) lists ten families of neognathous birds with the same configuration of three tympanic recesses and adjacent pneumatic spaces. Saiff (1974, 1976, 1978a,b, 1981, 1982, 1983, 1988) published a series of papers on the middle ear of birds [including Procellariiformes (1974), Sphenisciformes (1976), Pelecaniformes, and Ciconiiformes (1978b), and paleognaths (1978a, 1981, 1982, 1983, 1988)]. He recognized only two tympanic recesses, the "upper tympanic recess," comprising the superior and posterior tympanic recess, and the "presphenoid sinus," which is synonymous with the anterior tympanic recess. His description is based on studies of the openings in the bony skull, and thus contains no detailed information about the anatomy of the recesses themselves. Norberg (1978) gives a detailed description of the three tympanic recesses in a very thorough paper on the functional anatomy of skull asymmetry and ear structures in Tengmalm's owl (Aegolius funereus). Despite variations based on the skull's asymmetry, the topography of the three tympanic recesses in Tengmalm's owl is very similar to that in the description just given. Macke (1969) found all three tympanic recesses in the skull development of the coot, *Fulica atra*. He did not see a fusion of the contralateral anterior tympanic recesses. However, this was based on an investigation of young embryos, in which the anterior tympanic recesses were not yet properly developed. Goldschmid (1972 a–c) encountered the same problem in his study of skull development in mousebirds. He refers to the rec. tympanicus anterior as a blind air sac, emerging from the tympanic cavity and intruding into the basal plate up to the hypophyseal pit. He studied embryos and young nestlings (3 days old) in which an interaural pathway was not yet established.

Chatterjee (1991) recognized the same pattern of tympanic recesses in the triassic *Protoavis texensis*, probably the oldest known true bird. Witmer (1990) studied the craniofacial air sac system of mesozoic birds and found openings for anterior, superior, and posterior tympanic recesses in all mesozoic birds, including Archaeopteryx. Furthermore, he found that the contralateral anterior tympanic recesses communicated in the parasphenoidal bone as described here for modern birds. This pattern is also found in theropods (e.g., *Troodon formosus*; Currie 1985). Therefore, the presence of three tympanic recesses and of an interaural pathway are shared characters in all recent birds and were probably already present in their nonavian ancestors.

The interaural pathway has received some attention from physiologists as a structure providing a cue as to directionality in the periphery of the avian hearing system. This is an imporant function since the typically small head of birds does not allow them to gain directional information from acoustic sources by measuring interaural time differences or using intensity differences in the acoustic shadow. The interaural pathway might function as a sound pressure gradient receiver, which produces an effective directional intensity cue. Such a mechanism produces directional information through pressure gradients at the eardrum generated by sound pressure transmitted through the interaural pathway and transmitted around the head externally. Physiological results are controversial. On the basis of different experimental designs, Wada (1923), Schwartzkopff (1950, 1955, 1973), Payne (1971), Hill et al. (1980), Coles et al. (1980), Knudsen and Konishi (1979), Knudsen (1980), Rosowski and Saunders (1980), and Lewald (1990) found evidence that the interaural pathway may be of importance as a mechanism in sound localization in the domestic chicken, domestic quail, barn owl, and pigeon. However, Moiseff and Konishi (1981) showed that in the barn owl, sound attenuation in the interaural pathway is too large to enable the two ears to function as coupled sound pressure difference receivers, at least in the frequency range of 6–8 kHz, which is important for prey localization.

In conclusion, three tympanic recesses are a common feature of birds and are already present in mesozoic species. This is also true for the interaural pathway. There is evidence that the interaural pathway functions as a sound pressure gradient receiver in at least some species and provides a peripheral mechanism for directionality of the avian ear. The air volumes enclosed in the pneumatic spaces of the skull might also produce resonance effects as suggested here. However, physiological data to test this hypothesis are lacking.

3.7 Siphonium and Pneumatization of the Mandible and the Quadrate Bone

The quadrate bone and the mandible are pneumatized from the tympanic cavity. The quadrate is pneumatized through a window on the medioventral side of the otic process. A pneumatic duct connects the tympanic cavity with the mandible. In many species, the wall of this tunnel becomes cartilaginous or osseous with age and is then described as siphonium.

American Rhea. In the American rhea, the siphonium opens lateral to the tympanic opening of the pharyngotympanic tube. This position is dorsal and lateral to the origin of the paratubal muscle (see following). From there, the siphonium leads anteriorly, on the lateral side of the parasphenoidal bone. It appears as a tube between the paratubal muscle and the parasphenoidal bone. The siphonium runs through very loose connective tissue between the basicranium, quadrate, and mandible, and enters the internal process of the mandible on its lateral side (quadrate side), close to its internal tip.

The quadrate is pneumatized directly from the middle ear cavity. The pneumatic window is situated on the anterior edge of the otic process close to the quadratosquamosal articulation. In the hatchling, pneumatization has proceeded through the proximal part of the otic process only. Cavities in the distal part of the otic process and the corpus quadrati are still filled with marrow. Pneumatization later extends to these parts of the quadrate during postnatal ontogeny.

Crested Tinamou. Meckel's cartilage, the cartilaginous precursor of the articular bone, has not yet ossified in the hatchling; thus, pneumatization has not begun. However, a siphonium is already present and leads from the tympanic cavity to the internal process of the mandible. The siphonium opens lateral to the tympanic opening of the pharyngotympanic tube. From there it leads directly, anteriorly and ventrally, to the lateral side of the internal process of the mandible. It passes through connective tissue medial to the quadratomandibular articulation. In the hatchling, the siphonium leads to a shallow pit in the internal process of the mandible. At this position, chondroclastic cells are found to begin carving out the cartilage, indicating the origin of both ossification and pneumatization. However, at this developmental stage the siphonium ends there. In an adult, a well-developed pneumatic window is present at the medial side of the internal process of the mandible. All parts of the articular bone are pneumatized; however, the more anterior parts of the mandible are not.

The pneumatic opening of the quadrate is in the same topographic position as described in the rhea.

Emu. The siphonium opens in the farthest anterior corner of the middle ear cavity, where the stapedial muscle (see following) inserts into the tympanic membrane; however, it is located well above the opening of the pharyngotympanic tube. From that position, the siphonium leads anteriorly and ventrally through dense connective tissue into the internal process of the mandible, which has already ossified. The siphonium runs within the articular as an epithelial tube without any dilatations. Pneumatization has not yet started and the bone is still filled with marrow. The wall of the siphonium consists of dense connective tissue. Its inner surface is covered by a flattened epithelium of the same kind as in the tympanic cavity.

The pneumatization of the quadrate has already started. The pneumatic window is in the middle part of the ventral face of the otic process of the quadrate.

Cassowary. In the cassowary, the anatomy of the siphonium resembles that of the rhea. However, ossification of Meckel's cartilage has already begun and a thin layer of perichondral bone is found on the upper side of the cartilage. The siphonium does not enter the cartilage but ends at the position where the pneumatic window will later develop. In an adult, the pneumatic window is close to the medial tip of the internal process of the quadrate. As seen on CT scans, in adults, pneumatization reaches the anterior dermal bones of the lower jaw.

The quadrate of a cassowary hatchling is already pneumatized. Pneumatization enters through a lateral window on the otic process, close to the anterior border of the tympanic membrane. At hatching, most of the otic process, with exception of the part close to the quadratosquamosal articulation, as well as most of the body of the quadrate are pneumatized. Only a few cartilaginous areas are left. In an adult specimen, the whole quadrate is pneumatized and filled with a trabecular network. Serial CT scans have confirmed that the bony trabecular network correlates with pneumatization in the adult cassowary and not with bone marrow.

Kiwi. The siphonium is short and wide in the kiwi, leading from the anterior corner of the tympanic cavity straight ventrally to the internal process of the mandible. It is positioned posterior to the paratubal muscle. The lateral wall of the siphonium is attached to the tympanic membrane; thus, in histological sections the tympanic membrane appears to insert into the internal process of the mandible at that location. The articular bone is already pneumatized in a kiwi hatchling. The adult shows pneumatization of the articular but not of the other bones of the lower jaw.

In a kiwi hatchling, the quadrate is not yet pneumatized. It was not possible to determine a prospective pneumatic opening. However, in the adult, several small foramina are present close to the quadratosquamosal articulation. It was not possible to determine any air in the quadrate of an adult kiwi from X-ray CT scans. Thus, although foramina are present, pneumatization seems to be reduced.

Ostrich. The stage of ossification in the hatchling mandible is very similar to that in the crested tinamou. Only a few small vessels intrude into Meckel's cartilage and show chondroclastic activity. Thus, no pneumatization has yet developed in the mandible. However, the siphonium is well developed, leading from the lateral and anterior corner of the middle ear cavity to the internal process of the mandible. Laterally and ventrally, the tympanic opening of the siphonium is covered by the origin of the paratubal muscle. The siphonium is a wide tube through connective tissue medial to the quadratomandibular articulation. In contrast to that of the tinamou, it enters the internal process of the mandible and becomes completely enclosed by the cartilage of the internal process. From there, it continues posteriorly as a well-defined and rounded tube, very much as in the emu.

In an adult ostrich, a pneumatic window is located on the lateral side of the internal process of the articular bone. The extent of pneumatization is similar to that in the crested tinamou, in which only the articular bone is pneumatized, whereas all dermal elements of the mandible remain solid.

Compared to the rhea and the tinamou, no differences can be found in the pneumatization of the quadrate bone.

Barred Button Quail. In the hatchling barred button quail, the siphonium is a thin tube leading from the farthest anterior corner of the tympanic cavity to the internal process of the mandible. Its topographic relationship to the paratubal muscle, tympanic membrane, and lateral wing of parasphenoidal bone is similar to that of the American rhea. However, since Meckel's cartilage has not yet ossified at hatching, the siphonium does not enter the lower jaw.

A high degree of pneumatization characterizes the lower jaw of adult specimens. As seen from serial sections, pneumatization reaches far anteriorly into the dental, angular, supraangular, splenial, gonial, and articular.

Pneumatization of the quadrate has already begun at hatching. The pneumatic opening is present on the medial and anterior faces of the otic process. From there a large pneumatic recess enters the otic process of the quadrate.

Comparison. In comparison, the siphonium, pneumatic foramina, and pneumatization of the quadrate are highly similar in all species. The extent of pneumatization of the lower jaw varies, with the kiwi as an extreme in which no pneumatization is present, even in adults.

3.8 Pharyngotympanic Tube

The pharyngotympanic tube (tuba pharyngotympanica, tuba Eustachii) connects the pharynx and both middle ear cavities. Its anatomy is the same for all paleognathous birds, and will be described for the American rhea.

The pharyngotympanic tube opens into the anterior part of the tympanic cavity or, as a matter of definition, at the bottom of the anterior tympanic recess. It is medial to the origin of the paratubal muscle and the tympanic membrane. Beginning at the tympanic opening, the ventral side of the pharyngotympanic tube is covered by the lateral and ventral wing of the parasphenoidal bone. Thus, it runs through a bony canal on the ventral side of the basitemporal plate. In hatchlings, this canal is open, whereas in adult specimens it is closed, forming a bony tunnel on the ventral side of the basicranium. The length of this bony tunnel varies among species. The pharyngotympanic tube leads to the midline of the skull where it merges with that of the contralateral side to form a common tube. The common pharyngotympanic tube opens into the posterior region of the pharynx.

The epithelium of the inner wall of the pharyngotympanic tube is continuous with that of the tympanic cavity and the pharynx. In the lateral parts of the tube, the epithelium consists of a single layer of flattened cells on a lamina basalis. In the bony tunnel, the epithelial cells have a brush border, and the amount of mucoidal substance in the apical parts increases. In the common tube, there is a continuous transition of epithelium into the cuboidal to prismatic epithelium of the pharynx. The outer layer of the wall is a very thin sheet of connective tissue. In the bony tunnel this layer is tightly fused to the periost. In the common tube, the lamina propria is thicker and smooth muscle fibers are loosely interwoven. Close to the pharyngeal opening, tubular mucoidal glands are found in the deeper layer below the epithelium. The transition from the tympanic epithelium to the pharynx epithelium is a continuous change of cytological characteristics. However, there is no keratinization of the tubal epithelium as in the pharynx, and it is not possible to describe a tunica muscularis in the tubes.

The cross section of the tuba pharyngotympanica is compressed and no open lumen is present. Thus, in its "resting condition" the pharyngotympanic tube seems to be closed, very similarly to that in mammals. The few smooth muscle cells found in the lamina propria appear insufficient to open it. Instead, the adductor mandible muscles insert into the ventral side of the pharyngotympanic tube. The tube may be capable of opening through contractions of the adductor muscles when swallowing food. The mechanism is very similar to one described in mammals, although it involves different morphological elements (m. tensor veli palatini, Starck 1982).

In the barred button quail, muscle fibers of the tunica muscularis of the pharynx form a distinct muscle around the pharyngeal opening of the tube. This might function as an additional mechanism for closure of the pharyngotympanic tube.

3.9 Arthrology of the Quadrate Bone

American Rhea. The quadratocranial articulation is single-headed. The condylus has an ovoid form and extends over the squamosal (cotyla quadratica-squamosi) and the otic articulation facet (cotyla quadratica otici). The latter includes material from the crista parotica of the prootic bone and the metotic cartilage. A smooth indentation of the quadrate articulation head resembles the incisura intercapitularis of neognathous birds. The quadratocranial articulation is a synchondrosis (junctura cartilaginea; articulatio synchondrosi), which fixes the otic process of the quadrate to the neurocranium and restricts movements of the quadrate. The otic process of the quadrate is positioned close to the zygomatic process of the squamosum and the lateral wing of the parasphenoid, to which it is fixed by connective tissue, which can be ossified in the adult.

The quadratomandibular articulation forms a lateral and a medial condyle. The longitudinal axes of both condyles meet at an angle of almost 90°; they also incline at an angle of about 90° in the horizontal plane. The medial side of the mandibular process of the quadrate articulates with the pterygoid and the lateral side with the quadratojugal arc. These articulations are no different from those of other birds.

Crested Tinamou. The quadratocranial articulation (articulatio quadratosquamosootica) resembles that of the rhea and is also a synchondrosis. However, movements of the quadrate bone seem restricted not only by this synchondrosis but also by strings of connective tissue connecting the lateral wing of the parasphenoidal bone to the lateral surface of the otic process of the quadrate. Dense layers of connective tissue also connect the medial side of the zygomatic process to the lateral side of the otic process of the quadrate. They are supposed to limit the kinesis of the quadrate against the neurocranium.

Emu. The quadratocranial articulation resembles that of the other species: it is a synchondrosis limiting movements of the quadrate as just described.

Cassowary. The quadratocranial articulation is like that of the tinamou. Restrictions in quadrate kinesis arise from the synchondrotic articulation and ligaments spanning the quadrate, zygomatic process, and lateral wing of the parabasasphenoidal bone. Furthermore, the zygomatic process of the squamosal bone is elongated and extends along the otic process of the quadrate; it is so closely fitted to its anterior surface that it imposes additional restrictions on quadrate kinesis.

Kiwi. The otic process of the quadrate bone has a double-headed articulation with the neurocranium (squamosal and prootic bone). As in the other paleognathous birds, this articulation is a synchondrosis, restricting movements of the quadrate.

Ostrich. The quadratocranial articulation of the ostrich resembles that of the rhea. The quadratomandibular articulation has two elongated ovoid condyles on the quadrate and equivalent articulation facets on the mandibular side. As in the rhea, the medial condyle of the ostrich is well developed, whereas it is rather inconspicuous in the cassowary. Lateral and medial condyles are perpendicular to each other. The length axis of the medial condyle is oriented mediolaterally, but anteroposteriorly in the lateral condyle. The medial condyle lies distinctly below the lateral condyle and fits into a pit formed by the articular end of the mandible. The quadratojugal process is short and has a pit-like lateral articulation facet.

Barred Button Quail. In turnix, the quadratocranial articulation is dicondyle. The condyles are separated from each other and form two synovial cavities with diarthrotic joints. The axis of rotation, defined by a common axis through both joints, is approximately at a right angle to the length axis of the skull. Thus, diarthrotic joints, degrees of freedom for movements in the articulations, the surrounding connective tissue, and the presence of a postorbital ligament allow for active, coupled kinesis of the skull (double rhynchokinetic skull (Zusi 1984). No limitations on quadrate kinesis have been recognized in turnix.

The orbital process of the quadrate is slender and about as long as the otic process. The lateral quadratojugal process is short and has a lateral pit-like articulation.

Comparison. The arthrology of the quadrate bone is uniform in paleognathous birds. In all species, the otic process of the quadrate ends in a flat, ovoid condyle with shallow notches along its long sides. The notches are deep in the kiwi and almost absent in the ostrich. The length axis of the articulation inclines by ca. 45° over the length axis of the skull. The condyle is covered by cartilage and enclosed in only one synovial capsule. The cranial articulation facet extends from the squamosal bone dorsolaterally to the crista parotica and the metotic cartilage ventromedially; it is synchondrotic in all paleognathous species. To a different degree, ligaments fix the quadrate to the zygomatic process and the basiparasphenoid. The inclination of the condyle and the close fixation of the otic process to neighboring bones probably restrict the quadrate kinetics of the paleognathous skull. However, this does not imply that the paleognathous skull is a rigid construction. It suggests reduced kinetics of the skull but sustained flexibility in the flexion zones of the quadratojugal, pterygopalatinal, and nasal bars (for terminology, see Bühler 1981). However, this flexibility results in a passive "responsiveness" of the palate to mechanical strain during feeding rather than an active kinetic mechanism. The loss of the "postorbital ligament" in paleognathous birds indirectly supports this view of a lost active, coupled kinesis. Passive kinesis of the paleognathous skull has already been suggested by Hofer (1948). Later, Zusi (1984) supported this view with his analysis of the mechanical properties of avian skulls. He describes the paleognathous skull as central rhynchokinesis, and provides evidence for passive kinesis. Peters (1987) suggested a mechanical interpretation of the paleognathous palate, in which most of its structural peculiarities are regarded as specializations to reduce uncontrolled kinetics but retain flexibility.

Systematicists have paid some attention to the structure of the quadratocranial articulation. Lowe (1926) interpreted the monocondylous quadratocranial articulation as "failure to attain the perfect bird type" in paleognathous birds. More modern systematicists' language considers the single-headed, quadratosquamosal condition to be the ancestral archosaur condition, whereas a double-headed, quadratosquamosoprootic condition has been described as derived character. It was questionable whether the clearly monocondylous condition of paleognaths represents the archosaur ancestral condition (e.g., Cracraft 1974, 1981, 1986) or is secondarily monocondylous and derived from the dicondylous condition. Ontogenetic evidence (e.g., Müller 1961, 1963) and the notched margins of the quadrate condylus in adults suggest that it is derived from a dicondylous condition. Furthermore, it has recently been shown that Archaeopteryx was dicondylous. (Haubitz et al. 1988; Haubitz and Wellnhofer 1989). However, the triassic Protoavis had the monocondylous quadratosquamosal articulation as found in all other archosaurs (Chatterjee 1991). It therefore seems reasonable to conclude that the monocondylous quadratocranial articulation in paleognaths is derived from the dicondylous condition. This represents an apomorphy for paleognaths, especially in persistent combination with the synchondrotic articulation and the loss of the postorbital ligament.

3.10 Columella: Stapes and Extracolumella

The evolutionary history of the avian columella and the traditional homologies of stapes and extracolumella are summarized in section 1.5.1. This chapter will concentrate on a description of the columella and features that may be of significance in functional and phylogenetic interpretations. For descriptive purposes, the nomenclature of Gadow (1891) and Krause (1901) has been adopted. The columella is divided into the bony stapes, comprising the clipeolus and the stipes (shaft of the stapes), and the cartilaginous extracolumella with its three processes, proc. tympanicus, proc. infracolumellaris, and proc. supracolumellaris.

Of course, the columella plays a central role in the transfer function of the middle ear. Several earlier papers have suggested lever systems of the columella, which might contribute to the amplification function of the middle ear. The length of the stapes, the tympanic process of the extracolumella, the infracolumellar process, and the supracolumellar process have been measured bilaterally; the results are given in Fig. 44. The functional implications of the linear measurements for a lever system in a functional model of the middle ear will be discussed in section 4.2.

American Rhea. The clipeolus or stapedial footplate is an oval bony plate that inserts into the oval window and is fixed to it by the annular ligament (ligamentum annulare; Fig. 45). The perilymphatic side of the clipeolus is more or less plane. At its anterior margin, the footplate smoothly bulges into the scala vestibuli (Fig. 46). The stipes is a strut-like bone with a funnel-shaped insertion on the tympanic side of the footplate. The stipes is attached excentrally to the anterior part of the clipeolus (Fig. 30). Several openings pierce the funnel-shaped base of the stipes, and pneumatic recesses enter the bony parts of the columella. The form and position of the windows is irregular, but all face the posterior side. The anterior side of the stapes is solid without openings.

The stipes is round and narrow in cross section and its surface is smooth. As it approaches the extracolumella, the stipes becomes more oval in cross section. At the junction to the cartilaginous extracolumella, the cross section is highly oval (Fig. 46). This junction is supposed to function as a hinge. Micromanipulations showed that movements of the stapes towards the extracolumella are possible in that joint. Structural observations suggest that the oval form of the stapes-extracolumella junction restricts movement to a rotation around the long axis of the junction. The cross section of the cartilaginous extracolumella increases shortly distal to the intracolumellar hinge. At this position, all of its four processes emerge.

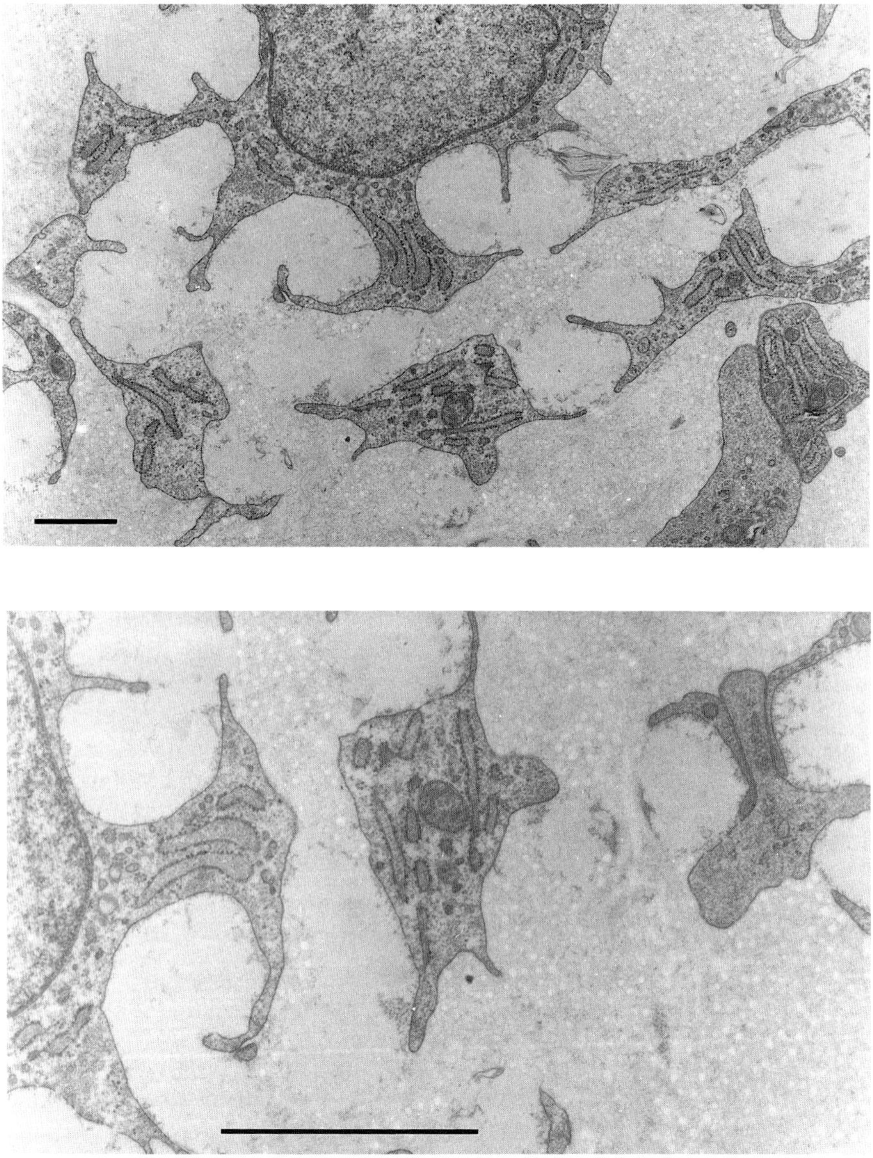

A

B

Fig. 44A,B. Domestic fowl, adult. Ultrastructure of Platner's ligament. **A** Cross section through the connective tissue core of the ligament, showing fibrocytes and an extended extracellular matrix; scale bar = 1 μm. **B** High-power magnification showing microfibrillar components in the extracellular matrix; scale bar = 1 μm

The tympanic process of the extracolumella (proc. tympanicus extracolumellaris) continues from the stipes as a straight cartilaginous rod. It inserts into the posterior part of the tympanic membrane where the ascending ligament (ligamentum ascendens) ends. The tympanic process is oval in cross section with a slight thickening at its tympanic insertion.

The superior process of the extracolumella (proc. supracolumellaris) emerges from the tympanic process near the intracolumellar hinge. It is irregular in cross section, and in the hatchling of the American rhea, it has a small foramen near its base (Fig. 19). This foramen separates the supracolumellar process into a medial and a lateral prong. The lateral prong is short and fuses with the medial prong at about half of its length. The medial prong inserts into the posterior margin of the tympanic membrane close to the bony margin of the tympanic membrane. The insertion of Platner's ligament close to the foramen suggests that it is equivalent to Huxley's foramen in neognathous birds. However, the foramen is not present in adult birds having only one supracolumellar process, the latter probably being a fusion of the lateral and medial prongs (Fig. 46).

The infracolumellar process (proc. infracolumellaris) emerges opposite the base of the supracolumellar process on the anterior side of the columella. It is the longest of all extracolumellar processes, emerging with a wide basis that continuously decreases in diameter during its course through the tympanic cavity towards the anteromedial margin of the tympanic membrane. The distal part of the infracolumellar process is embedded in the tympanic membrane. It ends just beneath the tympanic opening of the pharyngotympanic tube, but has no direct fixation to the bony frame of the tympanic membrane. A thin string of connective tissue rostrally connects the distal tip of the infracolumellar process with a small piece of cartilage, usually called the "stylohyal" cartilage. This stylohyal cartilage is embedded in loose connective tissue close to the lateral opening of the bony canal of the pharyngotympanic tube. In the American rhea, the stylohyal cartilage is well outside the tympanic cavity and has no direct connection to the tympanic membrane (Fig. 19). It was not possible to find it in an adult bird (Fig. 46).

Crested Tinamou. The clipeolus of the crested tinamou is flat and has a regular oval outline (Fig. 47). The stipes is centered in the clipeolus (Fig. 30). The basis of the stipes is broad and funnel-shaped. Pneumatic windows are present all around the basis. In its middle parts, the shaft of the stipes is round but becomes more oval distally, and is highly oval at the junction to the extracolumella. As in the American rhea, the intracolumellar junction serves as a hinge, allowing rotation only around its long axis.

The cartilaginous extracolumella has a wider diameter than the stipes. Three processes are present in tinamous (Figs. 21, 44). The tympanic process is the direct cartilaginous continuation of the stipes and abuts against the tympanic membrane.The knob-like tip of the tympanic process inserts into the end of the ascending ligament. The supracolumellar process emerges from the tympanic process, a short distance distal to the intracolumellar hinge. It points towards the posterolateral part of the tympanic membrane. The supracolumellar process has only a medial prong; a lateral prong is missing. Its distal end is thickened, with a hook pointing to the anteromedial side, which comes close to the ascending ligament. Assuming that medial and lateral prongs of the supracolumellar process represent the ancestral condition in paleognathous birds, the cartilaginous hook at the end of the supracolumellar process may be seen as the remnant of a lateral prong. However, since the

Dromaius novae-hollandiae

	left	right
Stipes	6.30 mm	
Proc. tympanicus	2.30 mm	
Proc. infracolumellaris	3.30 mm	
Proc. supracolumellaris	1.90 mm	
Force lever arm	6.04 mm	
Resistance lever arm	3.23 mm	
Lever ratio	1.86	

Apteryx australis

	left	right
Stipes		2.40 mm
Proc. tympanicus	1.29 mm	0.94 mm
Proc. infracolumellaris	1.70 mm	2.60 mm
Proc. supracolumellaris	1.36 mm	1.06 mm
Force lever arm	2.20 mm	2.70 mm
Resistance lever arm	1.80 mm	2.50 mm
Lever ratio	1.22	1.08

Casuarius bennetti

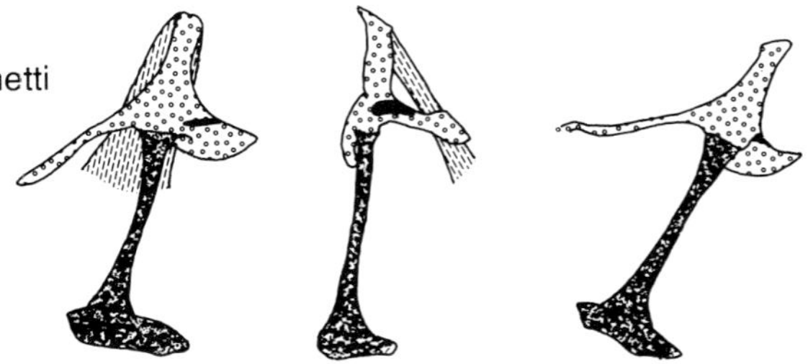

	left	right
Stipes	6.30 mm	5.90 mm
Proc. tympanicus	3.50 mm	3.70 mm
Proc. infracolumellaris	3.05 mm	4.20 mm
Proc. supracolumellaris	2.90 mm	2.80 mm
Force lever arm	5.50 mm	6.90 mm
Resistance lever arm	3.05 mm	5.00 mm
Lever ratio	1.79	1.38

Turnix suscitator

	left	right
Stipes	1.35 mm	1.40 mm
Proc. tympanicus	1.27 mm	1.30 mm
Proc. infracolumellaris	1.00 mm	1.10 mm
Proc. supracolumellaris	0.71 mm	1.0 mm
Force lever arm	1.81 mm	1.94 mm
Resistance lever arm	1.09 mm	1.07 mm
Lever ratio	1.66	1.81

Struthio camelus

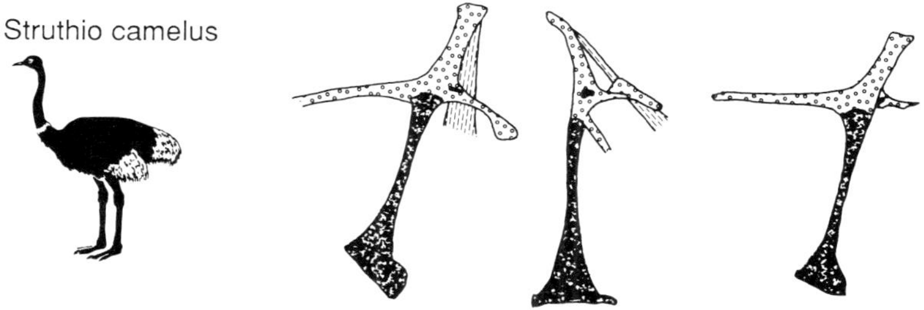

Linear measurements of columella and extracolumella

	left	right
Stipes		8.11 mm
Proc. tympanicus	5.56 mm	4.56 mm
Proc. infracolumellaris	5.60 mm	6.98 mm
Proc. supracolumellaris	3.50 mm	4.46 mm
Force lever arm	8.9 mm	7.15 mm
Resistance lever arm	5.4 mm	4.4 mm
Lever ratio	1.60	1.62

Rhea americana

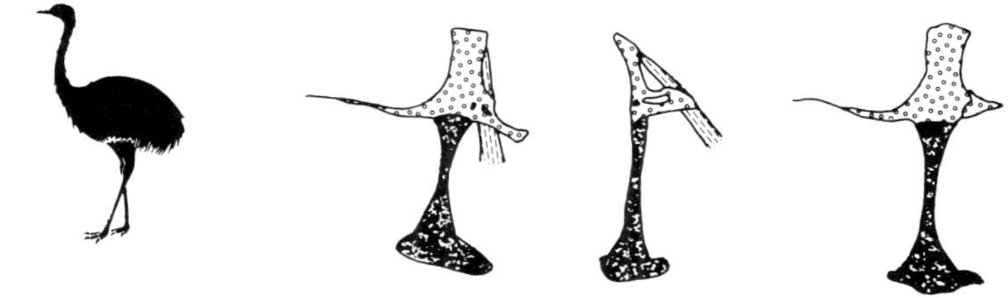

	left	right
Stipes	4.90 mm	4.86 mm
Proc. tympanicus	2.86 mm	3.15 mm
Proc. infracolumellaris	4.86 mm	4.34 mm
Proc. supracolumellaris	2.30 mm	3.09 mm
Force lever arm	5.40 mm	5.34 mm
Resistance lever arm	4.50 mm	4.34 mm
Lever ratio	1.20	1.23

Eudromia elegans

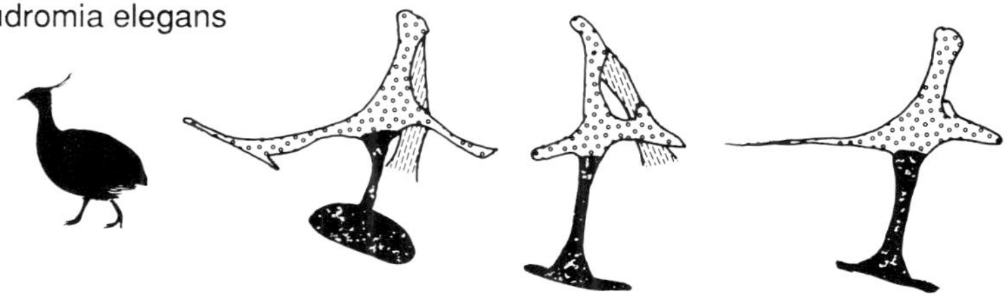

	left	right
Stipes	2.30 mm	1.88 mm
Proc. tympanicus	2.50 mm	1.75 mm
Proc. infracolumellaris	2.60 mm	2.60 mm
Proc. supracolumellaris	1.29 mm	1.33 mm
Force lever arm	4.00 mm	3.60 mm
Resistance lever arm	2.63 mm	2.64 mm
Lever ratio	1.52	1.34

Fig. 45. Topography and linear measurements of the stapes and the extracolumella. Columella is shown from medial, caudal, and lateral views. Linear measurements for left and right columella are given on the *right*

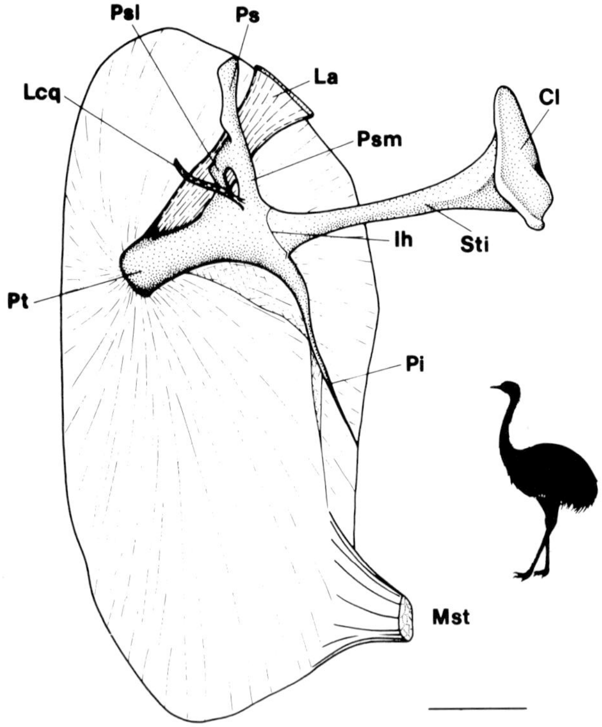

Fig. 46. American rhea, adult. Tympanic membrane, extracolumella, and stapes. Camera lucida drawing from dissection; dorsolateral view; scale bar = 2 mm

lateral prong of the supracolumellar process is reduced, no foramen of Huxley is formed.

The infracolumellar process in tinamous shows some peculiarities in comparison to the other species (Figs. 21, 44). Its broad basis extends from the intracolumellar joint at about half of the length of the tympanic process. From there it passes towards the anterior and medial margins of the tympanic membrane and then bifurcates. One branch continues the main course of the infracolumellar process along the margin of the eardrum and leads to its anterior and medial parts. The other branch bends backwards and follows the margin of the tympanic membrane in a posterolateral direction. Both branches of the infracolumellar process are embedded in loose connective tissue but contact the dense connective tissue (margo fibroelasticus), which fixes the tympanic membrane to the surrounding bones. The position of the infracolumellar process and its two branches suggests that it may function as a brace from the columella against the bony wall of the tympanic cavity. The branches of the infracolumellar process end at the margin of the tympanic membrane and do not pierce it. Separate stylohyal cartilage is not found in the crested tinamou.

For comparison, serial sections through the head of a hatchling *Tinamous tao* were studied and the same branching of the infracolumellar process was found.

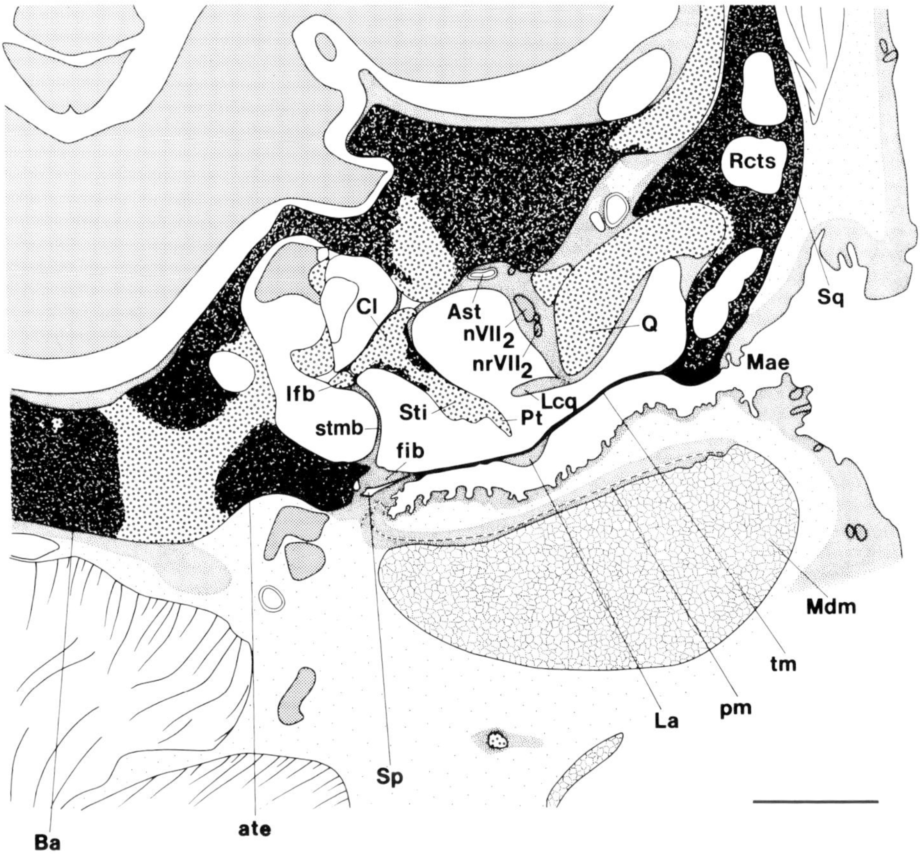

Fig. 47. American rhea, hatchling. Section through the left middle ear cavity showing stapes and clipeolus. Celloidin; serial section #398; 40 μm; azan; scale bar = 2 mm

Emu. Compared to other species, the emu has a simply structured clipeolus, stipes, and extracolumella (Figs. 22, 48). The clipeolus is a flat, bony, oval plate and does not bulge into the scala vestibuli. The stipes inserts centrally into the footplate with a slighty enlarged basis. Pneumatic windows open at the base of the stapes. The shaft of the stipes has a rounded cross section at its base, gradually changing into an oval cross section at the junction with the extracolumella. Three processes of the extracolumella can be described: (1) The tympanic process extends from the bony stipes to the tympanic membrane. It is short in comparison to the length of the stipes. In the hatchling, it bends slightly to the ventral side before it ends with a cartilaginous plate inserted into the tympanic membrane. (2) The supracolumellar process emerges close to the intracolumellar joint. It leads to the posterior part of the tympanic membrane. Its tympanic end is thickened and has a short hook flexed to the anterior side. As discussed in the tinamou, this hook may be seen as a remnant of the lateral prong of the supracolumellar process. (3) The infracolumellar process passes to the anteromedial side of the tympanic membrane, where it pierces the margin

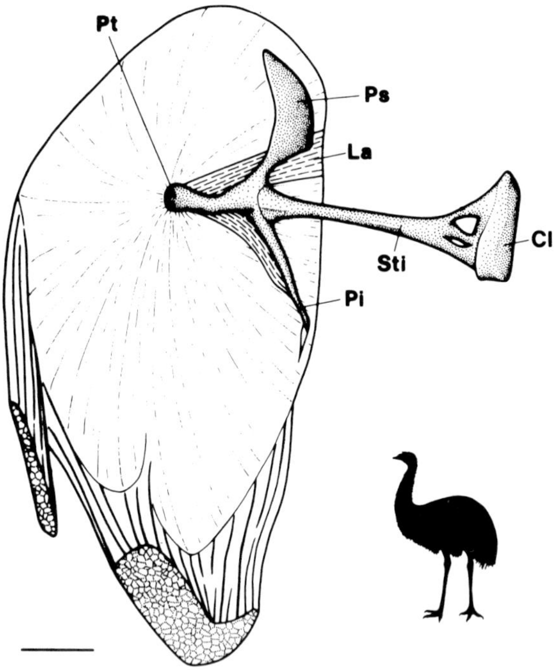

Fig. 48. Emu, adult. Tympanic membrane, extracolumella, and stapes. Camera lucida drawing from dissection; dorsolateral view; scale bar = 2 mm

of the tympanic membrane. It ends a short distance outside the tympanic cavity in a pocket formed by dense connective tissue. Within that pocket, the tip of the infracolumellar process is fixed only loosely so that it stays in position but can move. One might argue that this extratympanic part of the infracolumellar process represents stylohyal cartilage fused to the process. However, since we do not know the embryonic development of the process and since both the distal part of the infracolumellar process and stylohyal cartilage derive from the ceratohyal embryonic blasteme, it is impossible to confirm the presence of it as stylohyal in the emu.

Cassowary. In the dwarf cassowary and the southern cassowary, the perilymphatic side of the clipeolus bulges into the scala vestibuli, with a prominent hump on its anterior part. This hump is much more developed than in the American rhea. The posterior part of the perilymphatic side of the clipeolus is flat. On its tympanic side the footplate is flat, and the stipes inserts excentrally on the anterior part. The stipes shows no peculiarities in comparison to the other species. Although short in the hatchling (Fig. 24), it is a long and slender bone with a funnel-shaped basis in the adult cassowary (Fig. 44). Several pneumatic windows open on the posterior side of the stapes. The cross section of the middle part of the stipes is round. Near the intracolumellar hinge it becomes oval, and a hinge between stapes and extracolumella is formed as described for the other species. The tympanic process is a direct elongation of the stipes, thickening slightly where it inserts into the tympanic membrane. The ascending ligament connects it to the tympanic process as described previously (Fig. 44). The supracolumellar process emerges from the tympanic process near the intracolumellar hinge. It bends upward and posteriorly. Distally, it forms a cartilaginous triangle, with the long side of the triangle inserting into the tympanic membrane. Only the medial prong of the supracolumellar process is present; the lateral prong is probably reduced except for some material in the distal plate of the supracolumellar process. Consequently, Huxley's foramen is also lacking in this species.

The infracolumellar process emerges from the tympanic process just beneath the intracolumellar articulation and opposite the origin of the supracolumellar process. Its basis is broader than its diameter in a more distal position. The infracolumellar process points towards the anteromedial margin of the tympanic membrane. For a short distance it is embedded in the margin of the tympanic membrane and attached to its fibroelastic margin. Rostrally, the infracolumellar process leaves the tympanic cavity. It ends distally in connective tissue outside the middle ear, where it is tightly fixed to the bony frame of the tympanic cavity (basitemporal plate, lateral wing of the basisphenoid bone) by strips of dense connective tissue. The tip of the infracolumellar process comes close to the postmeatic membrane (membrana postmeatica). As discussed in the emu, the extratympanic part of the infracolumellar process might be interpreted as stylohyal cartilage. The position of the infracolumellar process, especially its connection to the bony frame of the tympanic cavity, and its tight fixation to the basitemporal plate, suggest that it might function as a lever arm in a system with the fixation to the basitemporal plate serving as a fulcrum (for a detailed discussion, see following). The fibers of an extraordinarily strong stapedial muscle extend to the distal part of the infracolumellar process, where they insert with a broad tendon. Such a direct insertion of the stapedial muscle into the infracolumellar process has not been observed in the other species.

The stapes of the cassowary has been previously described as being pierced by the stapedial artery (Freye 1952/1953). This observation cannot be confirmed on the basis of the material studied here. The pathway of the stapedial artery is identical to that in all other paleognathous species (section 3.15). Small openings on the base of the stapes are pneumatic windows through which recesses of the tympanic cavity enter the stapes.

Kiwi. The clipeolus of the kiwi has an oval outline and its perilymphatic surface is flat. The stipes inserts into the center of the tympanic surface of the clipeolus (Fig. 30). No pneumatic openings are found on the stapes. The cross section of the stipes is round but gradually changes distally into a more oval cross section. It is highly oval at the intracolumellar hinge.

The extracolumella has three processes, as described in the other species. However, the tympanic process is exceptionally short and thick (Fig. 25A,B). The supracolumellar process is a thin, cartilaginous branch directed towards the occipital and lateral side of the tympanic cavity. At its connection with the tympanic membrane, the supracolumellar process forms an enlarged fixation with the tympanic membrane. The lateral prong is lacking and consequently no foramen of Huxley can be found. In these respects, the supracolumellar process resembles that of the other species.

Origin and course of the infracolumellar process are similar to those in the other species. However, the process has a peculiar bend to the medial side. The distal end of the infracolumellar process is directed towards the marginal fixation of the tympanic membrane. It does not pierce the tympanic membrane. In the adult only, the infracolumellar process also has a discontinuity approximately at its midpoint. This discontinuity might be seen as a synchondrotic hinge in the infracolumellar process. Micromanipulations showed that it might serve as a flexion point if the distal end of the infracolumellar process is moved. However, we lack embryological data, and adult data are insufficient to determine the origin of this structure.

Ostrich. Clipeolus, stipes, and extracolumella of the ostrich (Fig. 49) are very similar to those in the emu. The perilymphatic surface of the clipeolus is completely flat. The funnel-shaped basis of the stipes is wide and inserts at almost the entire tympanic surface of the clipeolus. Pneumatic windows perforate the basis of the stapes. The shaft of the stapes is long and slender with a round cross section. Towards the junction with the extracolumella it

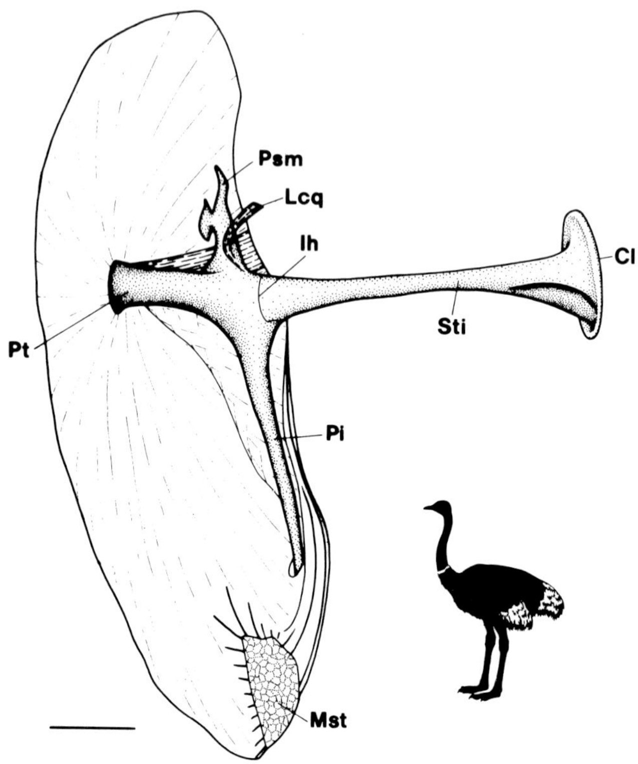

Fig. 49. Ostrich, adult. Tympanic membrane, extracolumella, and stapes. Camera lucida drawing from dissection; dorsolateral view; scale bar = 2 mm

broadens, but less so than is other species. However, micromanipulations showed that in the ostrich, too, the intracolumellar joint may function as a hinge.

The tympanic process is a continuation of the stapes and reaches the eardrum with a knob-like thickening where the ascending ligament inserts. The supracolumellar process emerges just distally to the intracolumellar junction with a broad basis. On its course to the occipital and lateral side of the tympanic membrane, it is thin and rounded in cross section but broadens in cross section close to the tympanic membrane. It inserts into the eardrum with an enlarged cartilaginous end. The supracolumellar process has no lateral prong and no foramen of Huxley (Fig. 28), thus resembling all other paleognathous birds except the American rhea. The infracolumellar process of the ostrich is long and slender, emerging opposite the supracolumellar process with a broad base, and then becoming thin and round in cross section. It leads to the anterior and medial margin of the tympanic membrane. There, it pierces the margin of the eardrum and leaves the middle ear cavity (Fig. 49). Its distal end, which may be called stylohyal (see preceding), is fixed into connective tissue medial to the depressor mandible muscle and the postmeatic membrane, and laterally to the internal carotid artery. The end of the process fits into a pocket formed by dense connective tissue. The infracolumellar process is fixed to that pocket by delicate ligaments that seem to hold it in position but also allow for movement. No direct fixation of the infracolumellar process to the bony frame of the tympanic cavity has been detected. On the contrary, the extratympanic part of the infracolumellar process is isolated from all surrounding solid structures. In the adult ostrich, a small tendon of the stapedial muscle inserts into the tympanic part of the infracolumellar process; this is not found in the hatchling.

Barred Button Quail. The columella of the neognathous barred button quail shows some features that are different from the paleognathous birds. Also, differences between the hatchling and adult are more distinct than in the paleognathous species. Therefore, adults and hatchlings will be described separately to account for ontogenetic changes.

Hatchling. In the hatchling, the clipeolus is still cartilaginous. It has an oval form. Its perilymphatic surface bulges smoothly into the scala vestibuli. The stipes arises from the center of the tympanic side of the clipeolus, with a broad cartilaginous origin (Fig. 29B). Perichondral ossification has deposited a thin layer of bone around the middle part of the stipes. Enchondral ossification has not yet occurred. At this stage of development, the cartilaginous stipes continues into the cartilaginous extracolumella. An intracolumellar hinge, as described for all the other species, has not yet been found.

The tympanic process (Fig. 29) is a prolongation of the stapes, but in contrast to the paleognathous birds, it bends somewhat to the anterior before reaching the eardrum. The supracolumellar process has two well-developed branches, a medial prong (proc. supracolumellaris medialis) and a lateral prong (proc. supracol. lateralis). The medial prong emerges distally from the (supposed) transition of stipes into extracolumella, whereas the lateral prong arises from the medial part at the tympanic process. Both prongs are directed occipitally and laterally and fuse at their distal ends. They enclose a large foramen of Huxley (Fig. 29A). The tympanic side of the lateral prong is fixed to the tympanic membrane along most of its length. The infracolumellar process of the extracolumella shows considerable difference compared to the paleognaths. It is the longest of the four processes and has an S-shaped form. From its origin on the tympanic process, it is a thin cartilaginous bar bending ventrally and anteriorly. After some distance it turns back in a ventral and lateral direction until it reached the ventral edge of the tympanic membrane. At this point, it changes direction again, now towards the anterior and in a horizontal plane. The distal and horizontal parts of the infracolumellar process are distinctly thicker than the more proximal parts. The infracolumellar process continues outside the tympanic cavity, almost up to the internal process of the mandible (proc. internus mandibulae) and the tympanic opening of the pharyngotympanic tube. No separate stylohyal cartilage has been found in turnix. However, in other neognathous species, the club-shaped part of the infracolumellar process is generally called stylohyal (see following). The extratympanic part of the infracolumellar process is tightly wrapped in the dense connective tissue of the ventral wall of the middle ear cavity. More rostrally, it is embedded in dense connective tissue and ends close to the internal process of the mandible. There, its tip comes close to a ventral sheet of the depressor mandible muscle but dose not touch it.

Adult. The clipeolus of an adult button quail has smooth perilymphatic and tympanic sides (Fig. 50). Its outline is oval but somewhat asymmetrical with one long side more rounded than the other. The stipes has a long but narrow, blade-like origin on the tympanic side of the clipeolus. The stipes is long, slender, and its cross section blade-like. At the transition to the cartilaginous extracolumella it has a highly oval cross section.

The tympanic process of the extracolumella is a straight elongation of the stipes. It has almost the same length as the stipes and points to the center of the tympanic membrane. The supracolumellar process is a thin hook, reaching the occipital portion of the eardrum. In contrast to the hatchling, there is only one branch and no foramen of Huxley in the adult. The hook-like form of the supracolumellar process implies a reduction of the lateral supracolumellar prong during postnatal development. The medial prong is still present and can easily be recognized by the insertion of Platner's ligament (Fig. 50). The infracolumellar process resembles that of the hatchling in its origin at the tympanic process and its S-shaped course through the middle ear cavity. However, considerable differences are present in its fixation to the basicranium. During postnatal development all the dense connective tissue ossifies, thereby closing the tympanic cavity ventrally and anteriorly (Fig. 23). Thus, the horizontal club-shaped part of the infracolumellar process, which is wrapped in connective tissue in the hatchling, becomes embedded in the bony floor of the middle ear cavity. Furthermore, the horizontal part of the infracolumellar process also ossifies but can be differentiated anatomically in adult specimens. The peculiar structure of the infracolumellar process and its incorporation into the bony ventral wall of the tympanic cavity is found in all button quail species.

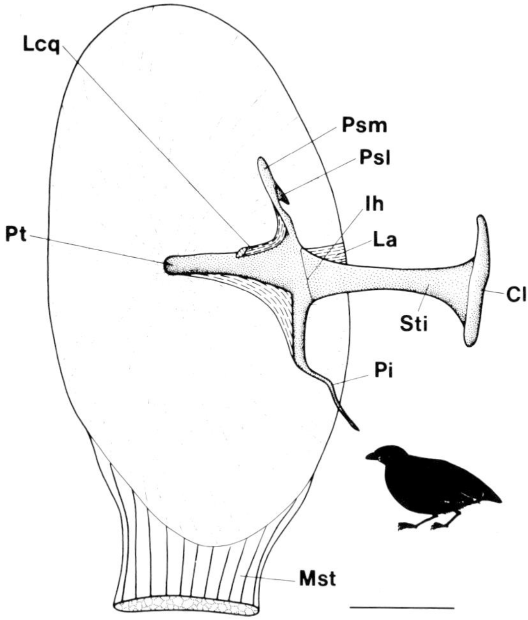

Fig. 50. Button quail, adult. Tympanic membrane, extracolumella, and stapes. Camera lucida drawing from dissection; dorsolateral view; scale bar = 1 mm

This specific structural arrangement has previously been described as the lateral basitemporal process (Bock and McEvey 1969).

Comparison. Comparative data for the avian columella (stapes and extracolumella) are scattered widely throughout the literature, with considerable variation in the quality of the data. Most descriptive craniological studies consider the columella an embryonic structure. The embryos investigated in such studies represent not only various species but also various ages and developmental stages. Since hatchlings and adults within one species show considerable differences (especially in turnix, but also in all other species investigated), such data must be used with caution in comparisons.

Krause (1901) described the structure of the bony stapes of 70 species and introduced the terminology currently in use. He reports a diversity in the forms of the clipeolus and the stipes, and in the number and position of pneumatic foramina. However, aside from a few extreme forms (e.g., in *Upupa epops* or *Tyto alba*), it is difficult to recognize characters that could be used for classification. The stapes of nonpasseriforme birds is relatively uniform. Feduccia (1974, 1975a,b, 1976, 1977, 1978) uses the morphology of the stapes to establish phylogenetic relationships for Passeriformes and related groups ("perching birds"). However, he recognizes derived characters in the stapes which could be used for phylogenetic classification only in the wood-hoopoes (Phoeniculidae) and hoopoes (Upupidae) on the one hand and bee-eaters (Meliphagidae) on the other. In most other taxa, the stapes has a rather plesiomorphic structure and resembles that described in the paleognathous birds.

Freye-Zumpfe (1952/1953) describes the stapes and extracolumella of 14 bird species. In all species, the clipeolus and stipes (shaft of the stapes) are very similar to that of the species described here and in other studies. Obviously, the stapes is rather uniform in birds and shows characterizing specializations only in the exceptions mentioned. Freye-Zumpfe (1952/1953) also provides comparative data about the cartilaginous extracolumella of adult specimens. In all species, she describes the extracolumella with its three processes in a manner very similar to the description given here for paleognathous species. However, she does not describe the attachment of the infracolumellar process to the wall of the tympanic cavity as found in this study. This might be a consequence of the exclusive use of macroscopic dissections, which are scarcely sufficient to detect an infracolumellar fixation. Huxley's foramen is not present in any of the adult specimens. This is particularily interesting since Huxley's foramen has been reported in the embryos of many species. It has been found in embryonic black-headed gull (*Larus ridibundus*) (Weber 1990) and in *Strix aluco* (May 1961) but not in adults of those species (Freye-Zumpfe 1952/1953). Brock (1937) and Frank (1954) mention a foramen of Huxley in early-stage ostrich embryos. However, aside from the data presented here, little other information is available for a comparison of embryonic and adult middle ears of the same species.

Many studies of cranial development in birds (see Table 1) also provide information about the extracolumella. It is obvious that the embryonic development of the extracolumella is essentially the same in all species. It has four processes, and the two prongs of the supracolumellar process enclose the foramen of Huxley. The foramen is not persistent and is reduced during posthatching ontogeny. Absence of the foramen in paleognathous birds may be due either to the relatively long incubation period of the species and the highly precocial developmental stage of their hatchlings, or to an actual lack of the foramen.

The infracolumellar process of the button quail has been described in detail as being characterized by a large, club-shaped horizontal portion embedded in dense connective tissue of the wall of the tympanic cavity. A similar club-shaped end of the infracolumellar process, often called stylohyal, was found in the great crested grebe, *Podiceps cristatus* (Toerien 1971), the coot, *Fulica atra* (Macke 1969), the nightjar, *Caprimulgus pectoralis* (Fourie 1955), the mouse-bird, *Colius indicus* (Goldschmid 1972c), the Japanese crane, *Grus japonensis* (Starck and Vorster, in preparation), and the pigeon, *Columba livia* (Starck, unpublished). Fourie (1955) was the only one to mention a similar attachment of the distal part of the infracolumellar process (stylohyal) to the margin of the tympanic wall; none of the other investigators paid attention to it. As shown by Frank and Smit (1976) and Smit and Frank (1979), the stylohyal cartilage is the most distal part of the embryonic hyoid arch and may be homologized with the distal portion of the ceratohyal element of the ancestral vertebrate condition. Considerable variability has been observed in birds, in that the stylohyal is either permanently separated from or permanently fused with the infracolumellar process, or both fuse during ontogeny (Macke 1969). Such variability has even been described in paleognathous birds, where only the rhea has a separate stylohyal. No stylohyal is found in the emu, cassowary, or ostrich, in which it might be fused with the infracolumellar process. It is entirely missing in the tinamou and the kiwi. At present, the pattern of variation observed in the stylohyal does not appear to be useful for phylogenetic analyses.

In conclusion, the avian stapes and extracolumella are characterized by their uniformity. In the adults of some taxa, there are peculiarities that may be used for phylogenetic analysis but also point towards interesting, mostly unresolved, diverse adaptations. The comparison of species faces the major problem that almost all information comes from individuals of different ages. Embryos are probably more similar to each other than adults in whom structures are ultimately differentiated, and whether hatchlings of altricial and precocial birds represent equal develomental stages (Starck 1989, 1993) remains unresolved. The fixation of the infracolumellar process to the wall of the middle ear cavity and the presence of an intracolumellar hinge are discussed as important functional features.

3.11 Ligaments of the Middle Ear

Two ligaments are associated with the avian middle ear and have been described in all species investigated so far. The ascending ligament (ligamentum ascendens) is a structure peculiar to the avian tympanic membrane. Platner's ligament (ligamentum columello-squamosum or preferably: ligamentum columello-quadrato) stretches between the intracolumellar hinge of the columella and the ventral edge of the otic process of the quadrate.

3.11.1 Ligamentum ascendens

American Rhea. The ascending ligament is a distinct fibroelastic ligament running through the middle layer of the tympanic membrane (Fig. 45). It is located in the mediooccipital region of the tympanic membrane, leading to the supracolumellar process and the tip of the tympanic process of the extracolumella. It is inclined at an angle of ca. 45° to the sagittal axis of the head (Fig. 6A,B). The ascending ligament originates from the ventrolateral part of the exoccipital bone, just beneath the opening of the glossopharyngeal (ninth) nerve. It enters the tympanic membrane directly behind the position where the hyomandibular nerve leaves it. Histologically, it consists of collagenous and/or elastic fibers. Smooth muscle cells have not been found in the ascending ligament.

Crested Tinamou. The ascending ligament in the crested tinamou is anatomically similar to that of the rhea (Figs. 12, 24). On external examination, it has the same position on the tympanic membrane and originates from the lateral and ventral edges of the exoccipital bone close to the foramina for the ninth and tenth nerves. It inserts at the tip of the tympanic process (Fig. 44). As seen in light microscopy, the ascending ligament consists mainly of collagenous fiber elements.

Emu. The ascending ligament of the emu is short and has a broad, fan-like origin on the lateral edge of the exoccipital bone. It passes crosswise behind the hyomandibular nerve and enters the middle layer of the tympanic membrane. Its collagenous fibers converge to form a string in the middle layer of the tympanic membrane. The ascending ligament inserts at the tympanic edge of the superior process and the tip of the tympanic process, where it ends (Fig. 48). Its inclination to the length axis of the skull is ca. 25° (Fig. 30). The ascending ligament consists of collagenous fibers; muscle fibers are not involved.

Cassowary. In the cassowary, the supracolumellar process of the extracolumella reaches far occipitally of the posterior margin of the tympanic membrane. Thus, the ascending ligament bridges only a very short distance between the lateral margin of the exoccipital and the posterior ends of the supracolumellar process. The short and stiff connection implies a firm fixation of the supracolumellar process to the exoccipital bone by means of the ascending ligament. From the superior process, the ligament leads to the tip of the tympanic process as in all the other species (Fig. 51).

Kiwi. In the kiwi, the ascending ligament is short (Figs. 48, 52). Origin, course, and insertion resemble those described for the emu. However, this ligament is not as prominent a structure in the kiwi; rather, it is a small local condensation of collagenous fibers.

Ostrich. In the ostrich, the ascending ligament is short, likewise originating from a ventrolateral position of the exoccipital bone, and runs into the tympanic membrane, where it inserts at the superior and tympanic processes of the extracolumella (Figs. 48–50). Its topography very much resembles that described in the emu.

Barred Button Quail. The anatomy of the ascending ligament in button quails is very similar to that of the paleognathous species (Figs. 48, 50). The ligament emerges from the lateral edge of the exoccipital bone in the posterior part of a notch between exoccipital bone and metotic cartilage. It originates in the posterior part of the tympanic membrane in close topographic relationship to the stapedial artery. It is a rather delicate structure and enters the tympanic cavity where the hyomandibular nerve leaves the cavity. It inserts at the tip of the tympanic process of the extracolumella.

Comparison. The ascending ligament is present in all birds. It is uniform with regard to both origin and insertion. It is a peculiar avian feature not found in any other sauropsid. "Homologization" with the ligamentum superius of lacertilians (Dombrowsky 1925; Freye-Zumpfe 1952/1953) is rather doubtful.

3.11.2 Ligamentum columello-quadratum (Platner's Ligament)

American Rhea. In the American rhea, Platner's ligament originates on the ventral edge of the otic process of the quadrate (Figs. 45, 46). This origin is

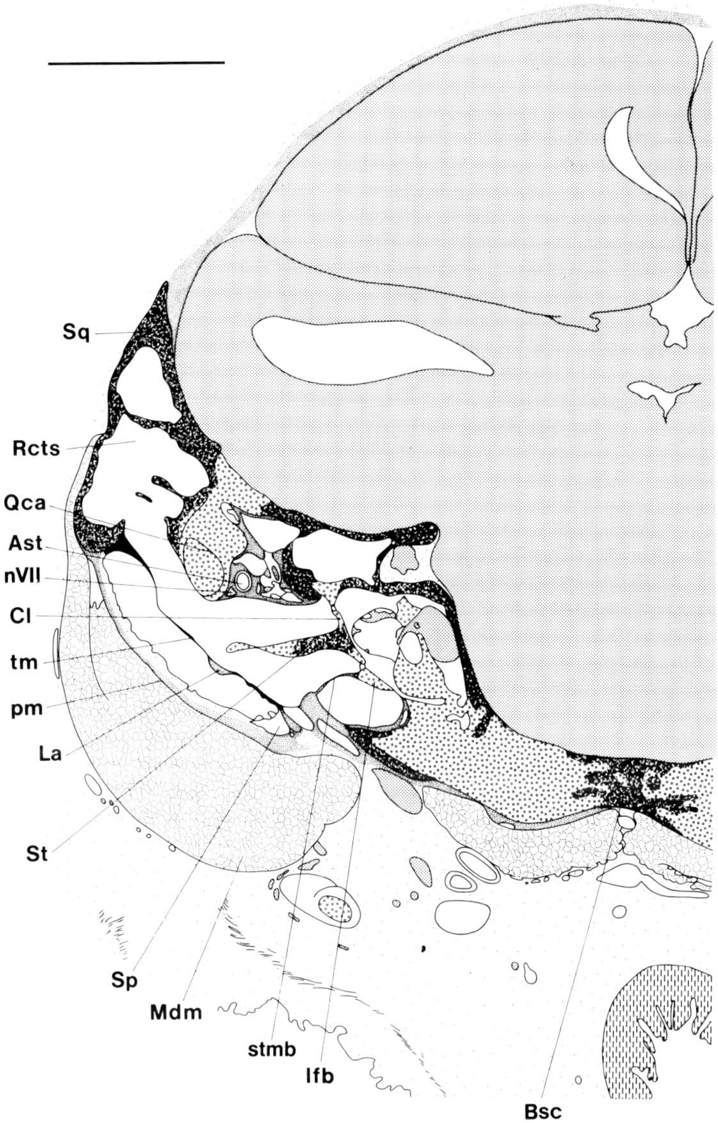

Fig. 51. Crested tinamou, hatchling. Camera lucida drawing, section through the right middle ear cavity showing stapes and clipeolus. Paraffin; serial section #61/6; 15 μm; azan; scale bar = 2 mm

clear and distinct, and it has no connection with the squamosal bone as suggested by the nomenclature. From its origin, the ligament crosses through the tympanic cavity and inserts into the extracolumella distal to the infracolumellar hinge on the basis of the supracolumellar process.

Crested Tinamou. Platner's ligament has the same origin, course, and insertion as described in the rhea. It originates from the ventral edge of the otic pro-

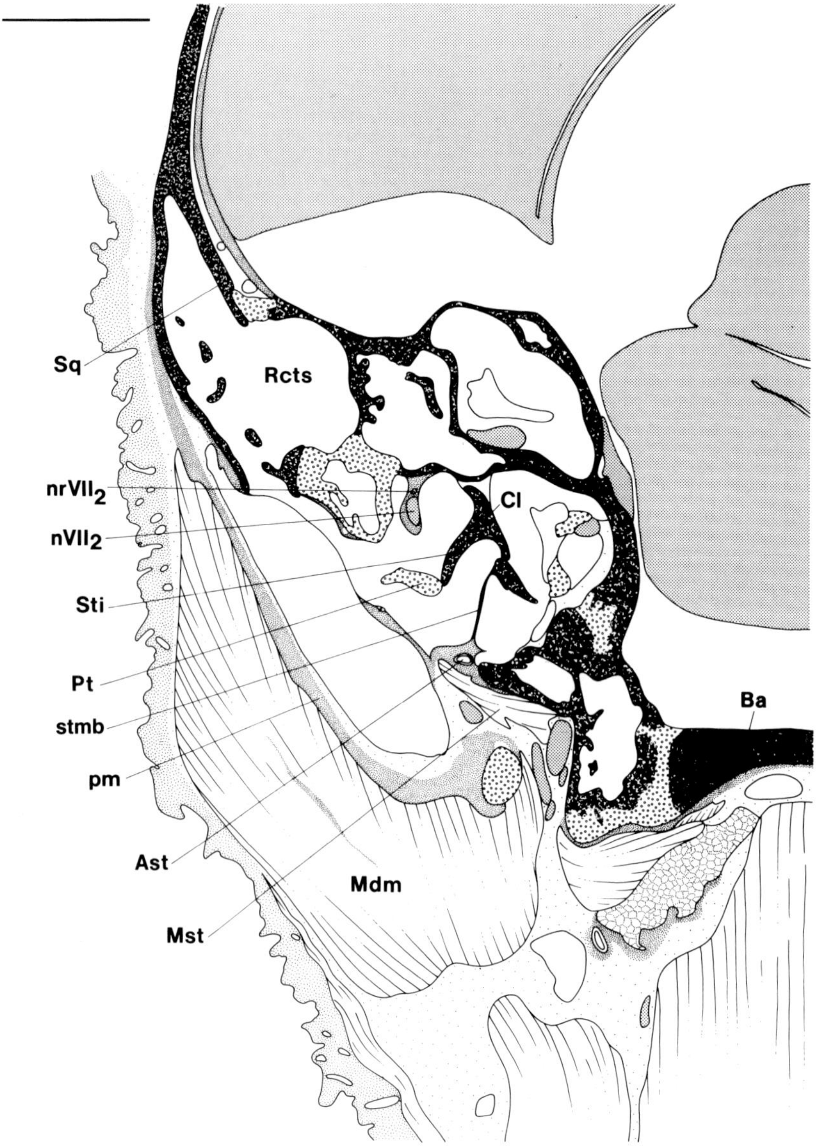

Fig. 52. Kiwi, hatchling. Camera lucida drawing of a section through the right middle ear cavity showing stapes and clipeolus. Celloidin; serial section #223/1; 40 μm, azan, scale bar = 2 mm

cess of the quadrate, travels through the middle ear cavity, and inserts on the tympanic process of extracolumella where the supracolumellar process branches off.

Emu. Origin, course, and insertion of Platner's ligament in the emu (Fig. 48) resemble those in the rhea. However, since the tympanic cavity of the emu is narrower than in the rhea and the tympanic process comes closer to the otic process of the quadrate, the ligament is shorter than in the other species. The ligament has an epithelial cover and a core of collagenous fibers.

Cassowary. Platner's ligament in the southern cassowary (Fig. 51) and the dwarf cassowary revealed no differences from the other species described.

Kiwi. In the kiwi, Platner's ligament is very delicate and short. It emerges from the ventrolateral edge of the otic process of the quadrate. Though close to the articulation of the quadrate with the squamosal, the site of origin is clearly on the quadrate. It inserts at the tympanic process of the extracolumella close to the intracolumellar hinge.

Ostrich. In the ostrich, Platner's ligament shows some peculiarities. As in the rhea, it originates from the otic process of the quadrate (Fig. 53) and its insertion is at the infracolumellar hinge (Fig. 49). However, the tympanic membrane of the ostrich forms a large epithelial fold in the anterior part of the middle ear cavity. This tympanic fold reaches the columella. Platner's ligament is cytologically recognizable as a distinct bundle of collagenous fibers embedded in the epithelial fold of the tympanic membrane (see 3.5).

Barred Button Quail. In the barred button quail, platner's ligament originates from the ventral edge of the otic process of the quadrate. It travels through the tympanic cavity and inserts at the tympanic process of the extracolumella, in a position where the supracolumellar process branches off from the tympanic process (Fig. 50). The course of the chorda tympani (see 3.13) is associated with the ligament.

Electron microscopy of Platner's ligament in domestic fowl (*Gallus gallus*) showed that the ligament is covered by a single-layered epithelium resembling the internal epithelium of the eardrum. A basal lamina is well developed. The "core" of the ligament consists of fibrocytes with a high portion of extracellular matrix. In electron micrographs, a predominant portion of the matrix has an amorphous component. Microfibrillar elements are clearly visible (Fig. 44). Thus, the ultrastructural findings reveal all elements of elastic tissue and make it probable that Platner's ligament has considerable compliance. Contractile elements have not been found. This supports earlier reports by Pohlmann (1921) who described Platner's ligament as being composed mainly of elastic tissue. Because of the rarity of material, no ultrastructural data regarding this ligament are available for paleognathous birds. However, it seems highly probable that here, too, it has the same composition as in the chicken.

Comparison. Judging by the data from earlier reports (Smith 1905; Pohlmann 1921; Freye-Zumpfe 1952/1953; Zumpfe 1952/1953) and the data presented here, the anatomy of Platner's ligament is uniform in birds. Platner's ligament has been "homologized" with the internal process of lacertilian columella (Dombrowsky 1925). However, this statement was based on a superficial similarity between the position of the ligament and the process. The phylogenetic distance between lacertilian sauropsids and birds makes it is highly improbable that the internal process is directly homologous with Platner's ligament.

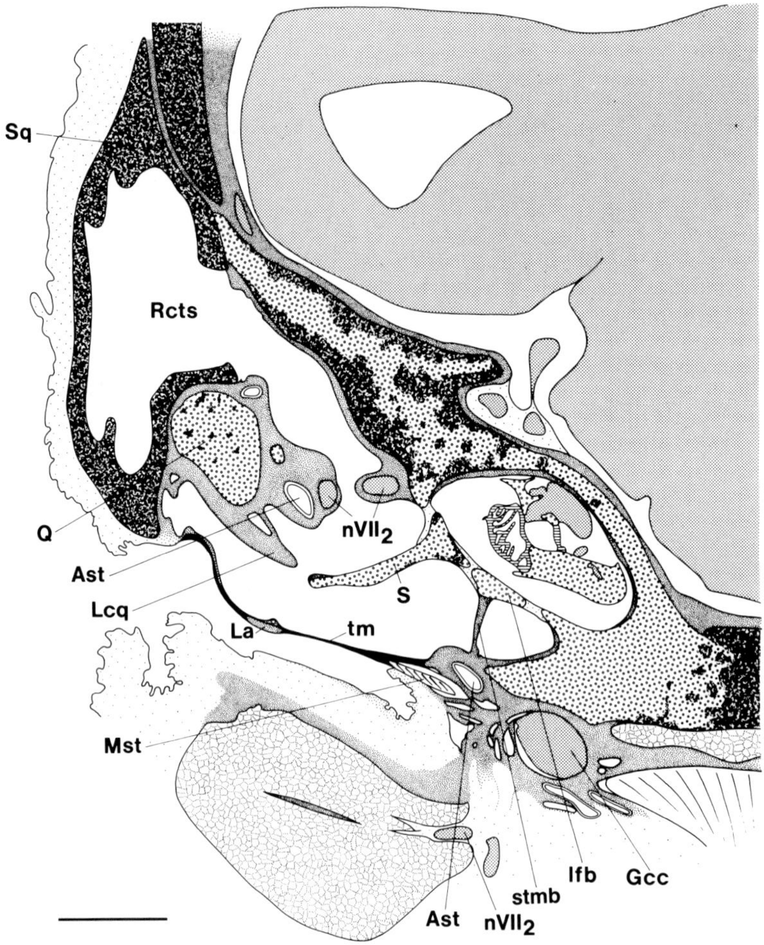

Fig. 53. Ostrich, hatchling. Camera lucida drawing of a section through the right middle ear cavity showing stapes and clipeolus. Celloidin; serial section #263/1; 40 μm; azan; scale bar = 2 mm

Pohlmann (1921) described the origin of Platner's ligament as extending from the otic process of the quadrate to the squamosal bone. Therefore, he called it the columello-squamosal ligament. Such an extended origin cannot be confirmed; this ligament should therefore preferably be named the ligamentum columello-quadratum.

No consistent hypothesis on the function of Platner's ligament is available. Suggestions concern its potential function as an antagonist to the stapedial muscle, to hold the stapes in place (Stresemann 1934) and provide additional stabilization (Kühne and Lewis 1985; Pohlmann 1921; Freye-Zumpfe 1952/1953; Norberg 1978). Recent functional morphological or physiological studies have not mentioned this ligament.

3.12 Muscles of the Middle Ear

Two muscles are recognized in relationship to the avian middle ear: the stapedial (musculus stapedialis) and the paratubal (m. paratubalis). Both insert into the tympanic membrane and not on the middle ear ossicle.

3.12.1 Musculus stapedialis

American Rhea. In the American rhea, the stapedial muscle has a broad and fleshy origin on the ventral side of the exoccipital bone, lateral to the opening for the combined ninth and tenth cranial nerves. The body of the muscle is a thin sheet of pinnate fibers, which converge to a relatively strong common tendon, laterally directed straight towards the tympanic membrane. It crosses ventral to the hyomandibular nerve and the stapedial artery, inserting into the tympanic membrane rostral to a position where the nerves leave and the artery enters the tympanic cavity. The tendon stretches through the middle layer of the tympanic membrane. It is not possible to assign a discrete point of insertion (Fig. 46). The stapedial muscle is innervated by a lateral branch of the hyomandibular nerve (see following).

Crested Tinamou. In the crested tinamou, the origin and insertion of the stapedial muscle resemble those described in the American rhea. As in the rhea, the muscle has pinnate fiber architecture. It inserts at the ventral margin of the tympanic membrane with a tendon which, however, cannot be traced within the tympanic membrane.

Emu. In the emu, the stapedial muscle has a fleshy origin from the ventral side of the exoccipital bone lateral and also slightly posterior to the jugal foramen. It does not cover the common exit of glossopharyngeal and vagal nerves as in the cassowary (see following). For most of its course, it runs between the medial and the lateral branches of the hyomandibular nerve. It has a pinnate fiber arrangement converging to a tendon. The tendon is short and muscle fibers enter the middle layer of the tympanic membrane (Fig. 48).

The stapedial muscle is innervated by a small nerve emerging from the lateral branch of the hyomandibular nerve. After branching off from its main stem, the nerve runs directly to the medial side of the stapedial muscle, where it enters it.

Cassowary. The stapedial muscle has a broad and fleshy origin from the ventral side of the exoccipital bone and an anterior part of the tympanic wing of the exoccipital. It covers the common foramen of glossopharyngeus and vagus nerves. Both nerves run between basioccipital plate and stapedial muscle for some distance after leaving the braincase. The fibers of the stapedial muscle converge to a central tendon, which inserts directly into the ventral part of the eardrum (Fig. 51). For most of its course, the stapedial muscle lies between the medial and the lateral branch of the hyomandibular nerve. It is innervated by small branches emerging posteriorly from the lateral hyomandibular nerve. (The rest of the lateral branch of hyomandibular nerve innervates the depressor mandible muscle).

Kiwi. Medial to the tympanic wings of the exoccipital bone, the basicranium of the kiwi has an additional set of small ventral and lateral wings. These basitemporal wings form a deep groove between the tympanic wings of exoccipital bone. The stapedial muscle originates on the lateral surface of the basitemporal wings (Fig. 53), lateral to the opening of the vagal nerve. On its course to the ventral edge of the tympanic membrane, the stapedial muscle passes through a small notch formed by the tympanic wing of the exoccipital and the basicranium. In comparison to the other species, the stapedial muscle in the kiwi is short and small. Its fibers converge to the insertion at the ventral margin of the tympanic membrane, but a pinnate structure cannot be described. Like the other species, it is innervated by a branch of the lateral hyomandibular nerve.

In the kiwi, the glossopharyngeal and vagal nerves leave the braincase on different routes. Thus, the stapedial muscle has a close topographic relationship only to the vagal nerve, not to the glossopharynegal nerve (see following).

Ostrich. In the ostrich, the stapedial muscle emerges from the ventral side of the exoccipital with a fleshy origin, lateral and also a little posterior to the jugal foramen. Origin, course, pinnate structure, and insertion resemble those in the emu (Figs. 49, 50).

Barred Button Quail. The stapedial muscle of the barred button quail shows no fundamental differences from that described in the paleognathous birds. It has a broad and fleshy origin on the ventral side of the exoccipital bone. All fibers of the stapedial muscle are directed towards the anterior side, where they insert on a common tendon; the muscle is pinnate. The posterior part of the muscle is lateral to the foramen of the vagal nerve but some of its posterior fibers cover the tenth cranial nerve. The anterior part is lateral and ventral to the vagal nerve, between the medial and the lateral branch of the hyomandibular nerve. The tendon of the stapedial muscle inserts on the ventral and lateral borders of the tympanic membrane (Fig. 50), just in front of the ascending ligament. Muscle fibers inserting into the tendon can be traced into the middle layer of the tympanic membrane. The tendon of the stapedial muscle runs through the middle layer of the eardrum until it ends, close to the distal tip of the tympanic process of the extracolumella.

Comparison and Functional Considerations. Pohlmann (1921) provides an extensive discussion of earlier papers on the comparative morphology and presumed "homologies" of the avian stapedial muscle. The avian stapedial muscle is sometimes considered "homologous" with a corresponding muscle by the same name in mammals. However, they have nothing in common but the innervation through a branch of the facial nerve. It would be more appropriate to consider both muscles as being independently derived from facial nerve-innervated musculature in their common stem group.

On the basis of our present knowledge, the anatomy of the stapedial muscle seems to be uniform in birds. Innervation of the stapedial muscle is provided by a small nerve emerging from the lateral branch of the hyomandibular nerve. The nerve enters the muscle in its anterior part from the lateral side. Wong et al. (1992) used retrograde labeling techniques to identify the motoneuron pool for the stapedial muscle. In domestic chicken they located a discrete group of 65 neurons along the dorsolateral border in the ventral subnucleus of the facial nuclear complex. Thus, experimental evidence supports the anatomical finding that the stapedial muscle belongs to the facial system.

The ultrastructure and contraction properties of the stapedial muscle in domestic chicken were investigated by Borg et al. (1979) and Counter (1978), who characterized it as fast-twitch muscle with high-ATPase activity and a single-endplate innervation pattern. Counter and Borg (1979) showed that stapedial muscle activity in the domestic chicken increases the impedance of the middle ear. It is regularly activated during the animal's own vocalization. However, very high external sound pressure could not excite contractions. Thus, they argued that the stapedial muscle has no specific acoustic reflex. Oeckinghaus (1978) and Oeckinghaus and Schwartzkopff (1983) investigated stapedial muscle activity in the European starling (*Sturnus vulgaris*) and showed that stapedius contractions during acoustic stimulation reduce the number of action potentials measured in the acoustic ganglion. They concluded that stapedial muscle activity acts as a band-pass filter, exerting protective effects against acoustic overstimulation, and that it attunes the ear's dynamic range to higher sound pressure levels. Oeckinghaus and Schwartzkopff (1983) experimentally stimulated an acoustic stapedial muscle reflex in the tawny owl (*Strix aluco*), confirming earlier results obtained by Golubeva (1972). However, the tawny owl is still the only species in which a stapedial muscle reflex was demonstrated.

3.12.2 Musculus paratubalis

American Rhea. The paratubal muscle was described by Müller (1963) in an American rhea hatchling. The muscle originates with fleshy fibers on the ventral face of the parasphenoidal bone. Its origin is close to the tympanic opening of the pharyngotympanic tube on the dorsolateral side of the lateral wing of the parasphenoidal bone (Fig. 34). The paratubal muscle leads occipitally and inserts into the anteromedial margin of the tympanic cavity. As previously described by Müller (1963), this muscle has two heads, one inserting into the tympanic membrane close to the opening of the siphonium, the other leading more occipitally and ending close to the distal tip of the infracolumellar process. This part of the muscle can be traced in the middle layer of the ear drum. However, it does not insert directly into the infracolumellar process. The paratubal muscle consists of smooth muscle fibers, and it is not possible to trace its innervation by standard light microscopy.

Crested Tinamou. A fully developed paratubal muscle does not exist in the crested tinamou. However, a delicate ligament can be found in the same position as the paratubal muscle in the rhea. It emerges from the lateral wall of the pharyngotympanic tube and from the lateral wing of the parasphenoidal bone. From there it leads occipitally into the tympanic membrane. It can be traced for a short distance, traveling in the middle layer of the eardrum.

Tinamous Tao. The situation is very similar in *Tinamous tao*. However, a few muscle fibers can be found around the tympanic opening of the pharyngotympanic tube, leading into the eardrum, where they can be traced as a very delicate ligament.

Emu. In the emu, the topography of the paratubal muscle is very similar to that of the rhea. Anteriorly, it originates as a thick plate of connective tissue from the ventral face of the parasphenoidal bone; more posteriorly, it emerges from the lateral wing of the parasphenoidal bone, lateral to the tympanic opening of the pharyngotympanic tube. As in the rhea, it has two points of insertion in the tympanic membrane: one close to the opening of the siphonium, the other with a long and thin ligament into the tympanic membrane. There is no connection to the infracolumellar process or any other structure of the extracolumella.

Cassowary. No paratubal muscle can be found in the cassowary. However, a tendon occupies the same origin and insertion as described for the paratubal muscle in the rhea and the emu. No muscle fibers have been detected in that tendon.

Kiwi. Because of the rather poor preservation of material, microscopic serial sections of the kiwi did not provide conclusive evidence as to whether smooth muscle fibers are present lateral to the opening of the pharyngotympanic tube. However, thick bundles of fibers were found in the same position as the paratubal muscle in the other species.

Ostrich. The paratubal muscle is well developed in the ostrich. It originates from the lateral side of the parasphenoidal bone. From there is runs posteriorly, traveling beneath the tympanic opening of the pharyngotympanic tube and entering the eardrum. The paratubal muscle is wrapped in a dense layer of connective tissue. Its insertion into the tympanic membrane is somewhat different from that in the other species because of the peculiar fold of the tympanic membrane. However, the paratubal muscle inserts into the eardrum and not into the fold. As already mentioned for the rhea, the muscle consists of smooth muscle cells and no innervation can be described.

Barred Button Quail. The barred button quail has a well-developed paratubal muscle. As in the other species described, the muscle originates from the lateral wing of the parasphenoidal bone and surrounding dense connective tissue, just lateral to the pharyngotympanic tube. The muscle leads directly into the anterior part of the tympanic membrane. Its fibers are well developed and the muscle is clearly recognizable. It is smaller and shorter than in the paleognaths. No ligament can be traced leading from the paratubal muscle into the tympanic membrane. As in all other species, it is impossible to recognize the innervation of the muscle.

Comparison. The paratubal muscle was first described by Müller (1963) in the rhea. A similar structure in the domestic chicken was recognized by Smith (1905), who described it as an extrastapedial ligament. Pohlmann (1921) recognized the same structure, and described it in greater detail as a superior and inferior drum-tubal ligament. Obviously, it is a regional differentiation of the wall of the pharyngotympanic tube, and its specific histology ranges from a pure elastic ligament (in the chicken, tinamou, and cassowary) to a solid bundle of smooth muscle (in the rhea, ostrich, emu, and turnix). Most probably it is derived from the tunica muscularis surrounding the pharyngotympanic tube. Since its innervation is not known, it is not possible to provide suggestions about its evolutionary derivation. Here, additional evidence is presented for all other paleognathous species and the neognathous turnix. Origin and insertion of the paratubal muscle suggest its function as a "tensor tympani." However, no data are available.

3.13 Nerves

Comparative morphology assigns the facial nerve (seventh cranial nerve) to the hyomandibular arch of ancestral vertebrates. In birds, the facial nerve forms a number of interconnections with the glossopharyngeal nerve; the vagal and the accessory nerve; the trigeminal nerve; and sympathetic as well as parasympathetic ganglia of the peripheral nerve system of the head. It will be necessary to discuss these interconnections and the peripheral ramifications of other cranial nerves for phylogenetic implications.

American Rhea (Fig. 19). The facial nerve leaves the braincase through a bony canal (canalis nervus facialis) in the prootic bone between the cochlear and the canalicular part of the auditory capsule. Immediately after leaving its canal, it forms the facial ganglion (ganglion geniculi), from where the facial nerve branches off into the palatine nerve (ramus palatinalis nervi facialis) and the hyomandibular nerve (ramus hyomandibularis nervi facialis).

The palatine nerve leads anteriorly and downwards on the lateral side of the prootic bone. It enters the anterior tympanic recess, which it crosses to reach the basisphenoid (Fig. 34). After a short distance it pierces the basisphenoid and enters the carotid canal through its own foramen. There, it fuses with a sympathetic nerve (n. sphenopalatinus, Müller 1963). The fused nerves form the vidian nerve (nervus vidianus). After a short distance, and before the parabasal canal turns dorsal, the vidian nerve splits off again, and the palatine part leaves the parabasal canal through a ventral foramen. From there it runs on the ventral side of the basitemporal process. The sphenopalatine part leaves the parabasal canal a little further anteriorly and travels on the dorsal side of the basipterygoidal process, together with the arteria sphenomaxillaris. It cannot be decided from standard light microscopy whether the palatine nerve and the sphenopalatine nerve exchange fibers during their common course as the vidian nerve. However, it is probable that they do, and that they have a different fiber composition when they leave the braincase.

The posterior branch of the facial nerve, the hyomandibular nerve, leads posteriorly and slightly dorsally to the ventral edge of the otic process of the quadrate (Fig. 47). From there it turns laterally and ventrally in a right angle, crossing the quadratocranial articulation and reaching the anterior edge of the metotic cartilage, where it makes a wide loop to the ventral side. It leaves the tympanic cavity at the occipital corner of the eardrum. The hyomandibular nerve splits into several branches. At the most caudal point of the hyomandibular slope, a small recurrent nerve (n. recurrens) branches off, turning back to the medial surface of the otic process of the quadrate on the same pathway as the hyomandibular nerve. It leads further to the anterior and connects with the supraorbital nerve of V_2 (ramus supraorbitalis nervi trigemini).

In the hatchling, an extremely thin chorda tympani branches off from the hyomandibular nerve (ramus auricularis) just behind its intersection with the ascending ligament but still within the tympanic cavity. From its origin, the chorda tympani follows the caudal margin of the tympanic membrane in a caudal and dorsal direction towards the metotic cartilage. On the lateral side of the metotic cartilage it runs dorsally to the quadratocranial articulation. After crossing this articulation, it follows a rostral direction on the medial side of the otic process. There it crosses lateral to Platner's ligament and reaches the otic process. It follows the otic process to a position close to the caudal end of the lateral wing of the parasphenoid, forming a bridge to the surface of the lateral wing of the parasphenoid. After a short distance it pierces the parasphenoid and reaches the ventral side. From there it runs to the lower jaw, along a pathway very similar to that in neognathous birds, except that it does not follow the columella as closely.

The chorda tympani of a hatchling rhea was described in detail by Müller (1963). Here, his observations can be confirmed. However, it must be pointed out that the chorda tympani of the hatchling is vestigial, extremely thin, and obsolete. I was able to trace it only on the left side of the head, not on the right side. Furthermore, it could not be found in the adult bird. Therefore, the presence of a chorda tympani in the hatchling American rhea must be considered with caution in phylogenetic considerations. A fully developed chorda tympani is not present in any of the other paleognathous species.

The hyomandibular nerve branches off into three nerves as soon as it leaves the tympanic cavity. The fibers of the lateral branch, ramus auricularis nervi facialis, extend to the skin of the external auditory meatus. This branch also sends a small nerve to the stapedial muscle, which it enters from the lateral side (see preceding). The major branch innervates the depressor mandible muscle. The medial branch leads anteriorly and connects to the cranial sympathetic ganglion (ganglion cervicale craniale sympathici).

The nervus glossopharyngeus (ninth cranial nerve) and nervus vagus (tenth cranial nerve) leave the braincase through a common foramen in the basitemporal plate. Shortly after leaving the braincase, the glossopharyngeus bends ventrally. The vagal nerve leads anteriorly and connects to the cranial cervical ganglion. The ninth and tenth nerves are directly connected to the cranial cervical ganglion; an exchange of fibers is probable. Another important connection between the cranial nerve system and sympathetic system is the n. sphenopalatinalis, a small sympathetic nerve leading anteriorly and joining the internal carotid, along which it enters the lateral opening of the carotid canal. In the carotid canal it lies lateral to the carotid. Later, it fuses with the palatine nerve to form the vidian nerve as described earlier. The vidian nerve splits off again after some distance. One branch, again called the sphenopalatine nerve (Müller 1963), leaves the carotid canal dorsal to the basitemporal process together with a small artery (arteria sphenopalatinalis). It then travels dorsal to the basitemporal process and leads anteriorly, connecting to the ophthalmic nerve. The ventral branch, ramus palatinalis n. VII (Müller 1963), leads ventral to the basipterygoidal process and connects anterior to the sphenopalatine ganglion.

The trigeminal nerve needs no detailed discussion here. However, the recurrent nerve of the hyomandibular nerve establishes an interconnection (Fig. 34) to the maxillary branch of the trigeminal nerve (ramus maxillaris nervi trigemini) before it ascends to the rete mirabile ophthalmicum. Facial fibers may also connect to the trigeminal system via the vidian nerve and fiber exchange with the sphenopalatine nerve.

Crested Tinamou (Fig. 21). In the tinamou, the course and arrangement of cranial nerves and peripheral nerves of the head is essentially the same as in the American rhea (Figs. 12, 25, 34). However, the chorda tympani is missing in the tinamou, even in early embryonic stages.

Emu (Fig. 22). The facial nerve anatomy of the emu resembles that of the American rhea. However, there are differences concerning the interconnections between facial nerves and the cranial sympathetic system. Two nerves emerge from the cranial cervical ganglion and connect with the palatine nerve. One leads more or less straight upwards (connexus ganglion cervicalis cranialis cum nervi palatinalis) and merges with the palatine nerve before it enters the parabasal canal. The second connection is with the sphenopalatine nerve, which enters the parabasal canal together with the internal carotid. It runs ventral to the internal carotid and extends to the dorsal palatine nerve within the parabasal canal. All three nerves exchange fibers extensively.

In all other repects, the anatomy of nerves in the tympanic region of the emu resembles that of the other species. The chorda tympani is not present in the emu.

Cassowary (Fig. 24). The anatomy of cranial nerves, their peripheral course, and their interconnections in the cassowary resemble those of the rhea (Fig. 51). The recurrent nerve of n. VII is covered by the stapedial artery (Fig. 54) and not shown in Fig. 24. The chorda tympani is lacking, as in all other paleognathous species except the rhea.

Kiwi (Fig. 25A,B). The topography of nerves in the ear region of the kiwi (Figs. 52, 55) is similar to that of the emu but also shows some unique features. The palatine nerve and the hyomandibular nerve emerge from the facial ganglion. The palatine nerve leads anteriorly into the anterior tympanic recess,

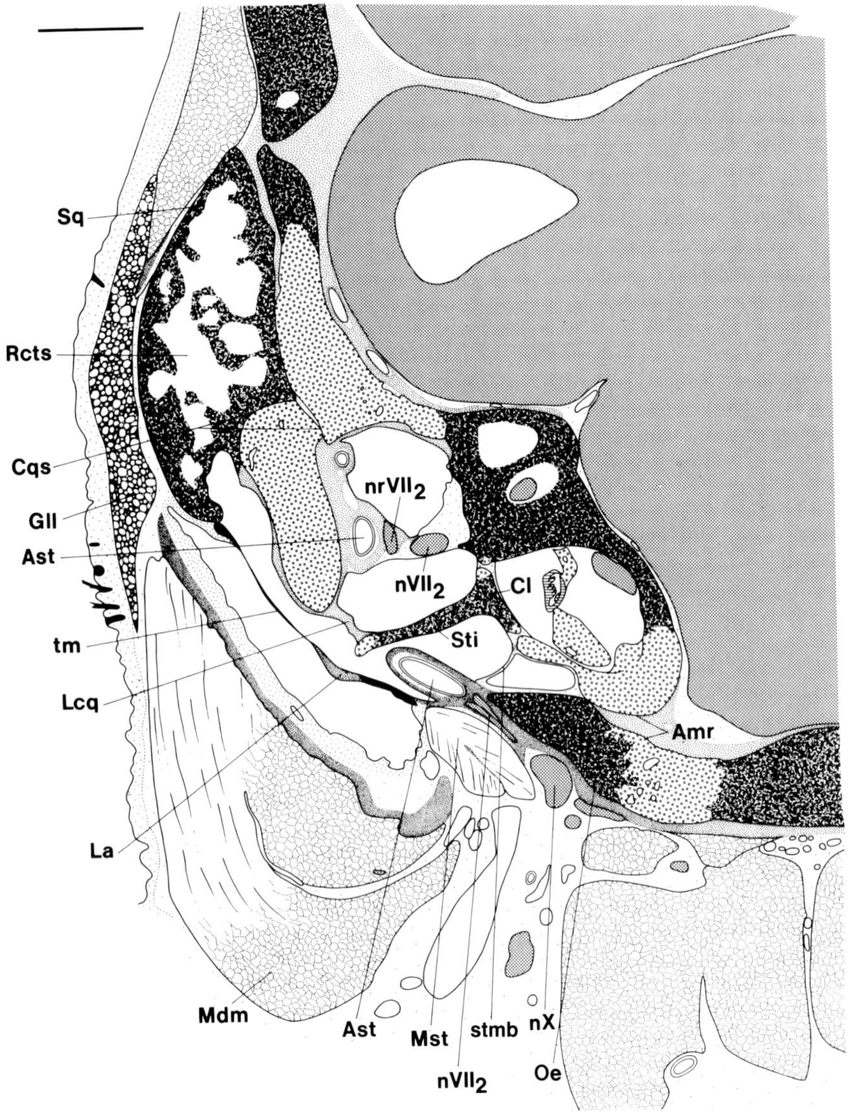

Fig. 54. Southern cassowary, hatchling. Camera lucida drawing of a section through the right middle ear cavity showing stapes and clipeolus. Celloidin; serial section #144/1; 40 μm; azan; scale bar = 2 mm

where it travels in rostral and ventral directions on the lateral side of the basisphenoid. The palatine nerve has a short and direct connection to the cervical cranial ganglion. This connecting nerve emerges close to the ganglion geniculi and leads straight ventrally to the cervical cranial ganglion. More anteriorly, the palatine nerve fuses with the sphenopalatine nerve for a short distance. In contrast to all other species investigated, the palatine nerve does not enter the carotid canal. Therefore, the vidian nerve lies outside the parabasal canal. However, as in other birds, the vidian nerve in the kiwi splits off

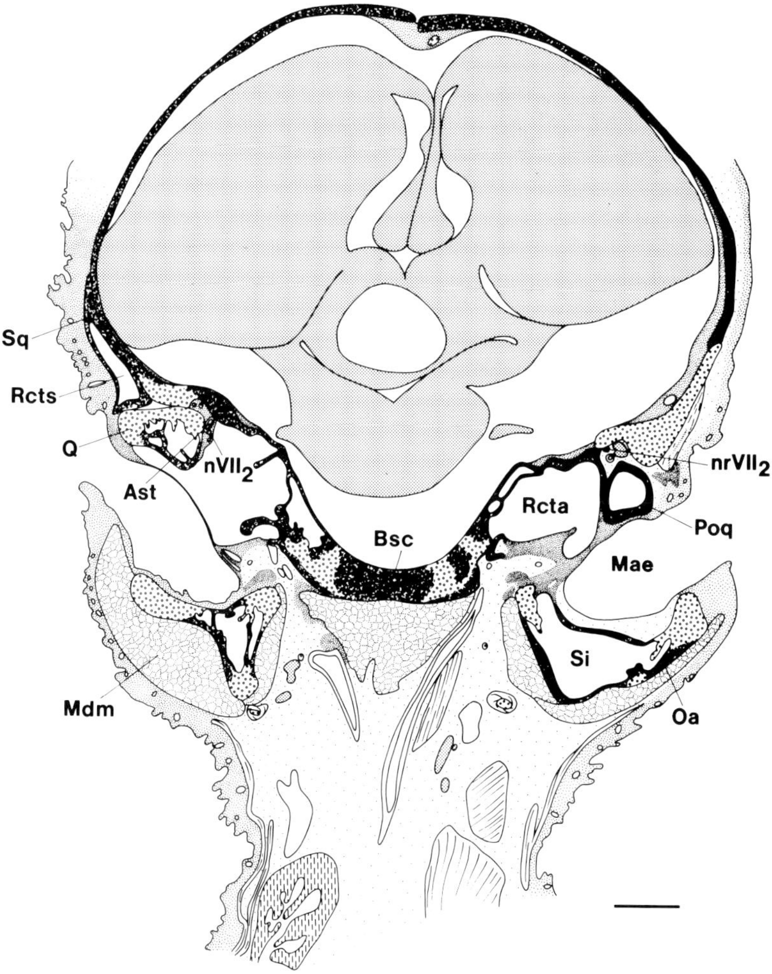

Fig. 55. Kiwi, hatchling. Cross section through the head. Camera lucida drawing of the anterior part of the tympanic cavity showing the opening of the anterior tympanic recess. Celloidin; serial section #195/1; 40 μm; azan; scale bar = 2 mm

anteriorly. The anterior part of the palatine nerve leaves the basisphenoid in a ventral position and travels ventral to the basitemporal process together with a small artery. The anterior course of the sphenopalatine nerve is above the basitemporal process together with the sphenopalatine artery.

The course of the hyomandibular nerve is similar to that of other paleognathous birds. However, the recurrent nerve (connexus cum n. V_2, Fig. 52) emerges directly caudal to the facial ganglion (Fig. 25A). From there it goes to the dorsal and medial side of the otic process to lead anteriorly where it connects to the maxillary branch of the trigeminal nerve (V_2). The peripheral branching pattern of the hyomandibular nerve is equivalent to that in other species. The chorda tympani is missing.

A major difference from other paleognaths concerns the exit of the ninth and tenth cranial nerves. In the kiwi, both nerves leave the braincase on different pathways. The glossopharyngeal nerve exits the braincase through the aquaeductus cochleae (= apertura medialis rec. scalae tympani) and passes through the recess of scala tympani (recessus scalae tympani), leaving the tympanic cavity through the foramen pseudorotundum below the membrana tympanica secundaria. The vagal nerve leaves the braincase through its own foramen in the basitemporal plate.

Several interconnections between the peripheral paths of the cranial nerves complicate the pattern. Portions of the hyomandibular nerve and the glossopharyngeal merge just where the glossopharyngeal leaves the tympanic cavity. The medial branch of the hyomandibular nerve connects with the cranial cervical ganglion. The hypoglossus nerve (12th cranial nerve) leaves the braincase through two foramina; a third branch leaves through the occipital foramen.

Ostrich (Fig. 49). Here again, the branching pattern and course of the facial nerve resemble those described in the American rhea (Figs. 41, 53). The chorda tympani is missing. Figure 49 also shows more distal details of the hyomandibular and vagal nerve connecting to the cranial cervical ganglion, where both nerves exchange fibers.

Barred Button Quail (Fig. 29). In turnix the facial nerve leaves the braincase through a canal anterior and dorsal to the oval window. There, it forms the ganglion geniculi and branches off into the palatine nerve and the hyomandibular nerve. The palatine nerve leads anteriorly, along the basisphenoid, and enters the anterior tympanic recess. A connection between the palatine nerve and the sympathetic sphenopalatine nerve is formed directly anterior to the facial ganglion. This connecting nerve pierces the basisphenoid in a separate canal. The palatine nerve leads anteriorly into the parabasal canal where it joins the sphenopalatine nerve to form the vidian nerve as described for the paleognathous species. The bifurcation of the vidian nerve and the anterior course of both resulting nerves below (n. palatinalis) and above (n. sphenopalatinalis) the basipterygoidal processes resemble those described in the emu.

The hyomandibular nerve slopes occipitally and ventrally parallel to the stapedial artery. The chorda tympani branches off from the hyomandibular nerve at its most caudal position. It crosses caudally to the stapedial artery, passing on to the tympanic membrane in an outward slope. Then, it travels along the inner side of the tympanic membrane to the tip of the supracolumellar process of the extracolumella, continuing on the dorsal side of the supracolumellar process to the insertion of Platner's ligament. There, it joins the ligament, traveling on to the ventral side of the otic process of the quadrate. Once it has reached the otic process it continues anteriorly until it bridges the gap to the lateral wing of the basisphenoid. Close to the basiparasphenoid it turns downward and reaches the articulation between the quadratomandibular articulation close to the siphonium. It enters a bony canal in the lower jaw (between Goniale and Meckel's cartilage) and joins the mandibular nerve. In its anterior course, the chorda tympani resembles that of other sauropsids.

In the specimen investigated, the chorda tympani of both sides of the head showed various courses. On the left side, the chorda differed from the preceding description by branching off further distally from the hyomandibular nerve, bending much further laterally, and meeting the supracolumellar process ventrally and in a more medial position. From there it passed through Huxley's foramen. Dorsally, it joined Platner's ligament as described for the right side. The anterior course of the chorda tympani resembled that of the left side of the head. This intraindividual variation should serve as a warning against far-reaching phylogenetic conclusions drawn on the basis of variations in nerve pathways. As pointed out for the chorda tympani of the American rhea, phenotypic plasticity is considerable, and one must discern between variability *within* individuals and variability among *individuals* of a species before drawing phylogenetic conclusions.

The distal branching pattern of the hyomandibular nerve resembles that described in the paleognathous species. A recurrent nerve was found to emerge in the same position as in paleognaths and with the same peripheral connection, where it merges with the supraorbital branch of the trigeminal maxillary nerve (n. $V_{2.1}$). Distally, the hyomandibular nerve divides into three branches with the same innervation pattern and peripheral nerve connections as described previously. A direct connection to the ninth cranial nerve is present (Fig. 29).

The glossopharyngeal nerve leaves the cranial cavity through the apertura medialis of the perilymphatic recess, follows a course directly below the perilymphatic sac on the basal plate, and passes laterally through the apertura lateralis of the perilymphatic recess (recessus scalae tympani). From there it bends ventrally, forming the distal glossopharyngeal ganglion and joining the cranial cervical ganglion. The vagal nerve leaves the braincase through its own foramen. Its fibers extend to the glossopharyngeal nerve (connexus vagoglossopharyngealis) and to the cranial cervical ganglion.

Comparison. Few studies have considered the peripheral anatomy of the avian cranial nerves, and the general representation is still confusing. However, an attempt will be made to summarize the common features of the facial nerve, the glossopharyngeal nerve, and the vagal nerve.

The position of the ganglion geniculi, the branching of the facial nerve into the palatine and the hyomandibular branch, and their peripheral course are identical in the avian species studied so far (domestic chicken, common swift, button quail, and paleognathous birds). In the rear of the middle ear cavity, the hyomandibular nerve splits off as a small recurrent nerve (n. recurrens) connecting to the n. V_2 (ramus maxillaris trigemini). The recurrent nerve is also found in domestic fowl (Cords 1904; Hsieh 1951; Nickel et al. 1977; Bubien-Waluszewska 1981), the common swift (Lorenz 1936),[2] and the Japanese crane (Starck and Vorster, in preparation). Although a confusing terminology complicates comparisons (see "List of Abbreviations and Synonyms"), the recurrent nerve is a structure common to birds.

[2] Historical note: Konrad Lorenz published this paper as a graduate student at the University of Vienna, some time before he started behavioral research.

A chorda tympani has been found in all neognathous birds [for a discussion of early papers see Cords (1904) and Bubien-Waluszewska (1981)]. It branches off from the hyomandibular nerve in its most caudal position. Its course through the tympanic cavity may vary between species but there is also considerable variation within species. The absence of the chorda tympani is a shared derived character in paleognathous birds. The description of a vestigial chorda in the right middle ear of the American rhea has been discussed as an ontogenetic and individual variation. Brock (1937) erroneously described the recurrent nerve as chorda tympani in an early-stage ostrich embryo. This was discussed in detail by Frank (1954), Webb (1957), and Müller (1963). Saiff (1982) recognized a chorda tympani in the kiwi, but probably also confused it with the recurrent nerve. Absence of the chorda tympani relates to a reduced tongue in all paleognathous birds (Bock and Bühler 1990) and a number of taste receptors and pharyngeal salivary glands that is probably reduced as well.

On its peripheral course, the hyomandibular nerve branches off into three branches. The lateral branch (ramus auricularis) innervates the external auditory meatus and the tympanic membrane. The main branch leads ventrally, where a small branch extends towards the stapedial muscle, before continuing to the depressor mandible muscle. The medial branch connects to the cranial cervical ganglion.

The course of the palatine nerve and its connection with the sympathetic sphenopalatine nerve to form the vidian nerve is a common feature in birds (Cords 1904; Lorenz 1936; Müller 1963; Nickel et al. 1977; Bubien-Waluszewska 1981; Starck and Vorster, in preparation). However, in the emu and kiwi, an additional connection between the palatine nerve and the cervical ganglion has been found. Turnix also shows an additional connection between both nerves; however, this connection is in a different position.

Phylogenetic relevance has been assigned to the exit of the vagal and glossopharyngeal nerves and to the branching of the facial nerve. In all neognathous birds, the ninth and tenth cranial nerves leave the braincase through separate canals. As described in detail for turnix, the ninth cranial nerve leaves the braincase either through a canal parallel to the aquaeductus cochleae (domestic fowl; Werner 1958; Bubien-Waluszewska 1981) or through the aqueduct (turnix, grus). Both canals develop during ontogeny by separation of the embryonic fissura metotica through a cartilagenous bridge. In some species the separation is missing or incomplete, resulting in only one canal or distally merging canals, respectively (Werner 1958). Distally, thc nerve travels below the perilymphatic recess before it leaves the middle ear cavity. The tenth nerve leaves the braincase through a separate foramen. This pattern is found in Galliformes, Anseriformes, Pelecaniformes, Gruiformes, Ciconiiformes, Psittaciformes, and Passeriformes (Werner 1958), as well as in Procellariiformes and Sphenisciformes (Saiff 1988). Because all other sauropsids show the same pattern (unpublished observation), it can be recognized as plesiomorphy.

In all paleognathous birds except the kiwi, the ninth and tenth cranial nerves leave the braincase through a common foramen. In the serial sections investigated here, the kiwi shows the pattern described in neognathous birds, with the ninth cranial nerve leaving through the aquaeductus cerebri and below the perilymphatic recess. In contrast, Saiff (1982, 1988) reports a common foramen for both nerves as in other paleognathous species. In all species the glossopharyngeus forms a ganglion (ganglion distale n. glossopharyngei) close to the cranial cervical ganglion. In some species both ganglia come so close to each other that they seem to fuse; in others a vagoglossopharyngeal connection is formed.

On the basis of the anatomical data reported here, the common exit of the ninth and tenth cranial nerves can be considered a character all paleognathous birds except the kiwi have in common. The kiwi follows a pattern found in neognathous birds and all other sauropsids.

3.14 Paratympanic Organ

The paratympanic organ (organon vitali) is a small vesicle-like sensory organ in the middle ear cavity of birds. The sensory epithelium consists of hair cells and supporting cells (von Bartheld 1990). Therefore, it has often been related to the vestibular or lateral line systems. However, the paratympanic organ is morphologically innervated by the facial nerve.

Paleognathous Birds. It was not possible to determine the paratympanic organ in paleognathous birds. Careful inspection of serial sections of the middle ear region of all species did not provide any positive evidence for the organ. Although its minuteness might make it difficult to detect it in 40-μm sections, it is probably missing in paleognathous birds.

Barred Button Quail. The paratympanic organ is well developed in the barred button quail. It is located a short distance behind the quadratocranial articulation in the dense connective tissue surrounding the stapedial artery. This organ is an elongated vesicle with a length of 470 μm and width of 150 μm (Fig. 56). These measurements correspond to comparative data, ranging between 350–1100 μm in length and 180–670 μm in width (Freye 1952/1953).

In the histological sections, the paratympanic organ is innervated by a small branch from the facial nerve. However, as described above, it is probable that the paratympanic nerve uses the facial nerve as a peripheral pathway without having a close functional relationship to it. It is not possible to resolve this question with standard light microscopy.

Comparison. The paratympanic organ is known from studies on many neognathous species (Oldenstam 1925; Freye 1952/1953). Recent investigations (von Bartheld 1990) have shown that paratympanic nerve fibers enter the brainstem together with the facial nerve but proceed to vestibular brainstem nuclei, with central projections similar to those of the lagenar nerve. The function of the paratympanic organ is still unresolved. These central projections suggest a function in equilibrium rather than audition. A function as air pressure receptor has also been discussed (von Bartheld and Rubel 1989). The paratympanic organ is probably an ancestral sense organ derived from the lateralis system of ancestral vertebrates. Its absence in paleognathous species

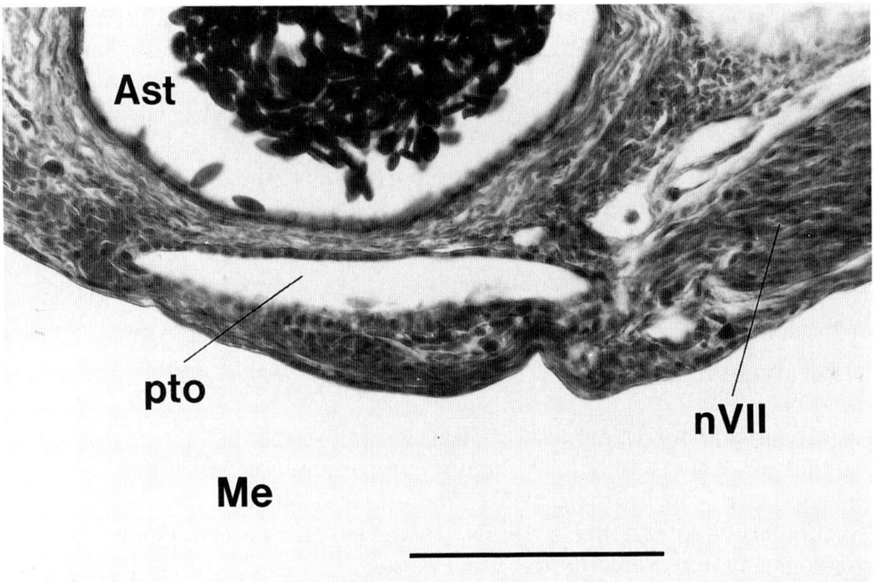

Fig. 56. Barred button quail, hatchling. Microphotograph of the paratympanic organ. Paraffin; 15 μm; serial section #36/1; azan; scale bar = 0.1 mm

must be considered a derived condition. Neognathous birds represent the ancestral condition.

3.15 Arteries

The vessels and blood supply of the avian middle ear are extremely uniform throughout all taxa examined. Therefore, the description given here covers all species investigated (Figs. 19, 21, 22, 24, 25, 28). Special attention will be paid to the course of the arteries.

The stapedial artery (arteria stapedialis = a. ophthalmica externa) branches off from the internal carotid (arteria carotis interna) before the internal carotid enters the carotid canal in the basisphenoidal bone. The stapedial artery slopes upward along the occipital border of the tympanic cavity and reaches the medial side of the quadratocranial articulation. It continues further anterior to the ophthalmic rete (rete mirabile ophthalmicum). There, it branches off into a number of small arterioles which are curled in the temporal fossa. In paleognathous species, the stapedial artery runs in a shallow bony groove. In the neognathous species (Fig. 29), it is covered by bone, thus running in a bony canal. As seen in the barred button quail, the bony cover of the stapedial artery is derived from connective tissue surrounding the vessel. Saiff (1988) describes a bony canal for the stapedial artery in paleognaths and many neognathous birds. As mentioned for turnix, the ventral cover of the stapedial artery ossifies from connective tissue. The considerable variability between individuals may cause these differences in findings.

Earlier descriptions of a stapedial artery piercing the stapes in the cassowary (Stresemann 1934; Freye-Zumpfe 1952/1953) cannot be confirmed. The course of the stapedial artery in the cassowary resembles that in all other birds. The foramina near the basis of the stapes are pneumatic openings.

The anterior branch of the internal carotid enters the carotid canal at the base of the skull between the parasphenoidal and basisphenoidal bones (Figs. 19, 34). Within the carotid canal, it is described as the cerebral carotid artery (a. carotis cerebralis). The palatine (a. palatinalis) and sphenopalatine arteries (a. sphenopalatinalis) bifurcate from the cerebral carotid, leaving the carotid canal through their own foramina together with the palatine and sphenopalatine nerves, respectively.

The cerebral carotids of both sides of the head connect in the upper part of the carotid canal. The intercarotidal anastomosis is established by a short horizontal vessel in the kiwi (Fig. 9), ostrich (Fig. 10), tinamou, emu, and button quail. In the cassowary and the American rhea (Fig. 6), both cerebral carotid arteries come close together and communicate in a side-to-side anastomosis. According to the terminology of Baumel and Gerchmann (1968), most paleognathous species and turnix show the H-type anastomosis, whereas the cassowary and the rhea have X-type anastomosis. It is rather doubtful whether these differences bear any systematic meaning as suggested by the abovementioned paper.

4 Discussion

4.1 Towards a Functional Analysis of the Avian Middle Ear

Section 1.4 summarizes the basic functional concepts of the avian middle ear. To recapitulate, the middle ear has been described as having an impedance matching or transfer function consisting of a *morphological amplification function* and a *morphological impedance*. Four parameters have been mentioned as contributing to the morphological amplification function: (1) the ratio of tympanic membrane and clipeolus, (2) the curved membrane effect of the eardrum, (3) A lever system formed by the columella and extracolumella, and (4) resonance of the air-filled spaces of the middle ear and its adjacent pneumatic spaces, which may contribute to a frequency-dependent amplification. The morphological impedance depends on: (1) mass reactance of all structures involved in the transfer function; (2) stiffness reactance, referring to the compliance of the system; and (3) mechanical friction.

4.1.1 Morphological Amplification Function

Area Ratio. The area of the tympanic membrane and the clipeolus were measured by planimetry. The area ratio was calculated by dividing the area of the eardrum by the area of the clipeolus. Results for the left and right ears of all species are given in Fig. 30. The area measurements of tympanic membranes range from 146 mm^2 in the ostrich to 13 mm^2 in the button quail; those for the clipeolus range from 4.7 mm^2 in the cassowary to 0.6 mm^2 in turnix. Area ratios between 37 in the American rhea and 22 in the button quail were calculated. There is some variability in the measurements of left and right middle ear areas, probably resulting from the small sample size and errors in measurement. However, the calculated values correspond well to a range of area ratios between 15 and 40 (Schwartzkopff 1955, 1957) for neognathous birds. Nevertheless, as mentioned in the introduction, physiological data suggest that only a part of the tympanic membrane is active and that the tympanic membrane is less responsive to sound transmission at high frequencies. Thus, investigation of the active area and frequency-dependent responsiveness of the tympanic membrane remain important tasks in functional studies of the avian middle ear.

Curved Membrane Effect. The effect of the curved membrane cannot be calculated from preserved material. As discussed by Tonndorf and Khanna (1970), the contribution of the curved membrane effect to the overall amplification function can be estimated by measuring the length between two marginal insertion points of the tympanic membrane and the maximum height of its excursion. Both measurements require live specimens and are meaningless on preserved material. Using laser interferography Tonndorf and Khanna (1970) determined a lever ratio of two for the curved membrane effect of the cat eardrum. No data on birds are available today.

Extracolumella Lever System. The extracolumella may contribute to the amplification function of the middle ear by acting as a lever system. On the basis of the anatomical studies presented in this paper, a lever system in the extracolumella can be suggested. The lever model is based on the observation that the infracolumellar process of the extracolumella is firmly but flexibly fixed to the wall of the middle ear cavity, and on the presence of an intracolumellar hinge in all species investigated. In this model, the stapes is not directly involved in the lever system but transduces the amplified oscillations directly to the clipeolus and the cochlear fluid. The lever system is located between the tip of the tympanic process of the extracolumella, the fixation of the infracolumellar process at the margin of the tympanic membrane, and the intracolumellar hinge. The fixation of the infracolumellar process to the wall of the tympanic cavity represents the fulcrum. For pressure transduction, the specific form of each of the processes is less important than the distances between the crucial points in the lever system. The straight line between the fulcrum and the tip of the tympanic process is considered to be a force lever arm, whereas the line between fulcrum and infracolumellar hinge is the resistance lever arm. Force lever arm and resistance lever arm rotate around the fulcrum. At the intracolumellar hinge the rotational movement is transformed into piston-like movements of the stapes and clipeolus (Fig. 57).

The length of the force lever arm and resistance lever arm has been measured as the distance between given points in microdissections, as just described. The results are given in Fig. 44. It must be mentioned that measurements of lever arms differ from linear measurements of the processes of the extracolumella. On the basis of the suggested lever system, amplification ratios between 1.1 (in the kiwi) and 1.8 (in the cassowary and button quail) were calculated.

Several other lever systems have been suggested in the literature (Pumphrey 1961; Gaudin 1968; Norberg 1978; Saunders 1985; Gummer et al. 1989a,b). Pumphrey (1961) and Gaudin (1968) considered the total columella complex to represent the force lever arm, with the clipeolus as the resistance lever arm, rocking asymmetrically around a hinge at the anterior margin of the oval window. Gaudin (1968) calculated a lever ratio of 4–5. However, this model does not match the structural data presented here. It does not consider the infracolumellar fulcrum and the intracolumellar hinge, which have been consistently demonstrated in all species studied here. The model also assumes an asymmetrical insertion of the footplate into the oval window that cannot be confirmed. Norberg (1978) based a lever system similar to that described here on his investigations of Tengmalm's owl (*Aegolius funereus*). Despite minor differences in the definition of the lever arms, the lever ratio of 1.6 he calculated

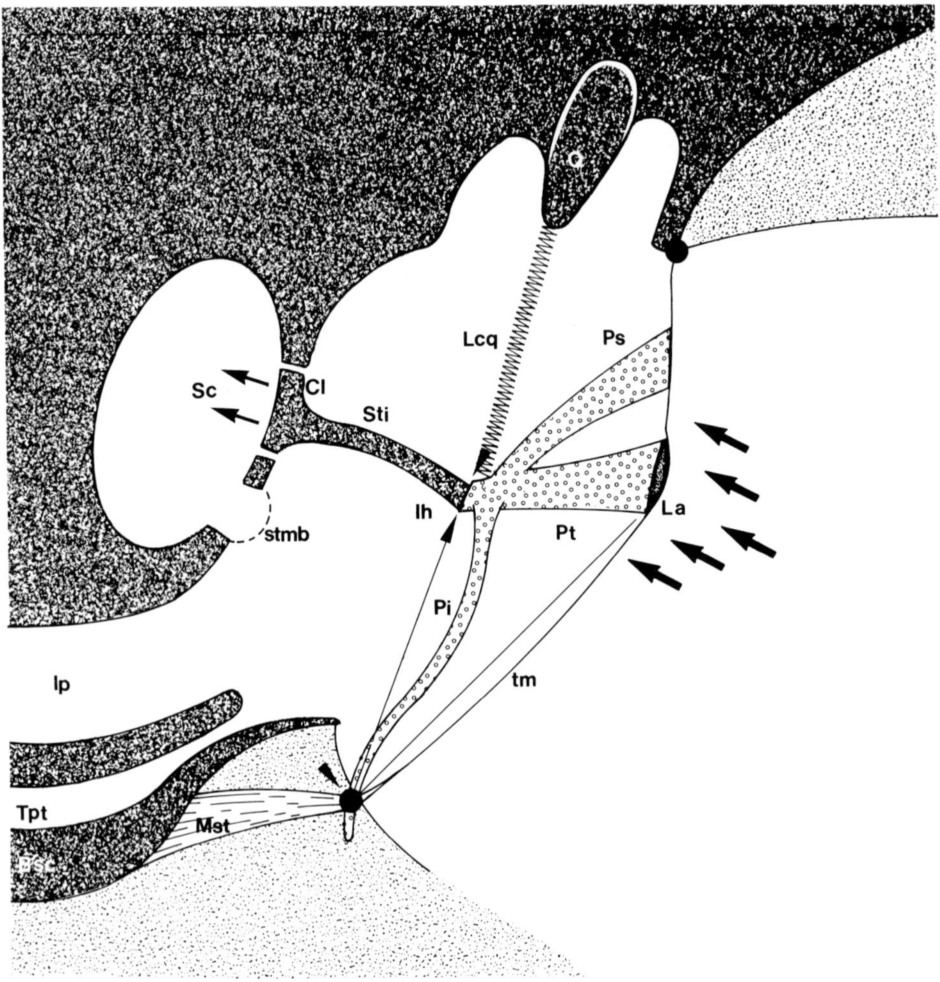

Fig. 57. Schematic drawing of the columella lever system as derived from structural findings. The extracolumella is thought to function as a lever system. The fixation of the distal end of the infracolumellar process to the margin of the tympanic membrane (•) or to the wall of the tympanic cavity is considered to function as a fulcrum (*arrowheads*). The *straight line* between the fulcrum and the distal tip of the tympanic process of the extracolumella represents the force lever arm, and a *straight line* between the fulcrum and the intracolumellar hinge is the resistance lever arm. Force lever arm and resistance lever arm rotate around the fulcrum. At the intracolumellar hinge, the rotational movement is transformed into pistonlike movements of the stapes and the clipeolus. Contraction of the stapedial muscle changes the mechanical impedance

for the owl lies within the range of data given here. Saunders (1985) avoided calculating a lever ratio on the basis of his morphological measurements. Gummer et al. (1989a,b) suggested an anatomical lever system similar to that of Norberg (1978) and the system developed here, and calculated a geometric lever ratio of 1.6 for the stapedial complex in the pigeon. However, the physiological input-output measurements they conducted in parallel ascertained an effective lever ratio of 2.7.

This discrepancy is a warning against conclusions based exclusively on structural investigations. While structural observations may be used to suggest functional models, any assumptions about functional effectiveness, e.g., of area ratios or lever systems, are hypotheses that need testing by independent physiological methods. Therefore, the data presented here should be considered as hypotheses, and not as the final results of functional investigations.

Resonance. The air volumes of the tympanic cavity and the adjacent pneumatic spaces contribute to an amplification function by frequency-dependent resonance (in addition to frequency-dependent resistance.) Resonance frequencies can roughly be approximated by assuming that the middle ear acts as a Helmholtz resonator. Large air volumes in the skull of the ostrich, cassowary, and American rhea result in very low resonance frequencies (Table 4), whereas the smaller air volumes in the head of the emu, tinamou, and button quail result in higher frequencies. The low resonance requencies suggest a low pass filter function. Low resonance frequencies might correlate with the low frequencies of the calls of most paleognathous birds. However, the large air cushion behind the eardrum raises the resilience of the middle ear. As mentioned previously, these models are speculative and should be considered preliminary hypotheses only. They must be replaced by more consistent functional models of the amplification function in the middle ear.

4.1.2 Morphological Impedance

Mass Reactance. The masses of all structures involved in sound transfer contribute to the mass reactance. Mass reactance increases with the mass of the structures involved in sound transfer, and the effects of mass reactance on the transfer function increase in proportion to frequency. Therefore, in an optimized system, masses would be expected to be reduced to the lowest possible values and adjusted to the frequency range of the species. Reliable data about mass of the tympanic membrane, columella mass, and mass of ligaments involved in the transfer function are still missing. However, without quantifying these effects and in a rather comparative anatomical sense, the pneumatization of such delicate structures as the stipes and clipeolus in all species except the kiwi can be interpreted as structural adaptation to reduce mass reactance.

Table 4. Volume measurements and estimated resonance frequencies

	Volume (mm^3)	Estimated resonance frequency (Hz)
Rhea americana	46634	776–908
Eudromia elegans	5478	1615–1753
Dromaius novae-hollandiae	40426	1151–1358
Casuarius bennetti	29876	321–375
Apteryx australis	8735	1279–1528
Struthio camelus	340349	307–355
Turnix suscitator	1560	3432–3707

Stiffness Reactance. Stiffness reactance describes the compliance of the system. It is controlled by its elastic tissue components and the air volume enclosed by the tympanic cavity. Stiffness reactance is high at low frequencies but rapidly loses impact on the transfer function as frequencies rise. Aside from relatively large air cushions in the paleognathous species, elastic tissue elements are found in the tympanic membrane, the ascending ligament, and Platner's ligament. In addition, the yielding properties of the extracolumellar cartilage and the stapedial bone contribute an as yet unknown proportion of the stiffness reactance. The tension of the eardrum might also affect stiffness reactance. Furthermore, the stiffness of the sound-transducing system is probably affected by the activity of the stapedius muscle. Although not yet shown in direct studies, it is highly probable that contractions of this muscle increase the tension of the tympanic membrane, thus increasing its stiffness and adjusting the ear's dynamic range to higher sound pressures (see 3.12.1). The role of the paratubal muscle is still unknown. Its topographic position suggests that its activity might also increase the stiffness of the sound-transducing system. It must be pointed out that an acoustic reflex, as reported in some mammalian species, is not generally found in birds (such as domestic fowl or the European starling). In general, the effect of the middle ear muscles is difficult to evaluate because of the unkown relationship between contraction strength and increased stiffness. Putting the few available data (see 3.12.1) into a functional context, it seems that changing the stiffness reactance through middle ear muscular activity is the only way to actively control the impedance of the middle ear and actively adjust its dynamic range to different sound pressure levels. In contrast, mass reactance and friction cannot be controlled actively.

Friction. Mechanical friction of all vibrating parts of the middle ear contributes to overall impedance. Friction is constant and independent of frequency. However, no data are available today.

Figure 57 graphically summarizes the functional implications of these structural investigations. As pointed out in section 1.3.3, this model is a static representation of those structures, which are considered functionally essential. The model represents a hypothesis about the mechanical function of the middle ear of paleognathous birds. Its validity can be tested by independent functional investigations of each of its components, and it can be improved by studying parameters such as material properties, which are inaccessible by structural analyses.

4.2 Phylogenetic Implications

Almost all structures of the paleognathous middle ear described and discussed here seem to be characters shared by all birds. Only a few structures are found to characterize paleognaths:

1. The pharyngotympanic tube runs through a short, bony tunnel in all paleognathous species. This tunnel develops between the ventral and lateral wings of the parasphenoidal bone after hatching. Saiff (1988) correctly recognized differences between the pharyngotympanic tubes of paleog-

nathous and neognathous birds. However, this tube is not naked as in his description; rather, it runs through a bony tunnel; the differences observed also concern the length of the bony tunnel. Cracraft (1974) discussed its significance for characterizing the neornithes. For the characterization of paleognaths he mentioned the pharyngeal opening of the tube far distal to the parasphenoidal rostrum. Strictly speaking, the cover of the pharyngotympanic tube is a weak phylogenetic argument since only gradual changes were observed in comparison to neognaths.

2. The quadratocranial articulation is monocondyle and synchondrotic in all paleognathous species. Notches at the margin of the condyle suggest its origin from a dicondylic condition. This view is supported by a dicondylic condition in all other birds, even Archaeopteryx (Haubitz et al. 1988; Haubitz and Wellnhofer 1989). In addition, the quadrate is characterized by ligaments fixing it to the skull. A long zygomatic process also restricts kinesis of the quadrate. The postorbital ligament, which in neognathous birds is related to cranial kinesis, is absent in paleognaths. This complex of characters is not found in other birds and may be considered an apomorphy of paleognaths. The data presented here reject Cracraft's (1974) view of this monocondylic articulation as ancestral.
3. A phylogenetically important structure is the common exit of the ninth and tenth cranial nerves, which is present only in paleognaths (except the kiwi). In all neognathous birds and in sauropsid outgroups, the ninth cranial nerve leaves the braincase through the aquaeductus cerebri and below the perilymphatic recess. The common exit for the ninth and tenth nerves is a derived character in all paleognaths except the kiwi. It is not possible to resolve whether the kiwi represents the ancestral condition or shows a secondary ancestral condition. However, the embryonic development of the exit of the ninth and tenth nerves shows that both leave the embryonic cranium through the large metotic fissure. During embryogenesis, the fissure is separated into the window for the tenth nerve and the medial opening for the perilymphatic recess. It is probably a question of embryonic brain size and width of the basal plate whether the ninth nerve runs through the first or the second window. The large skulls and relatively small brains of most paleognaths might cause the ninth and tenth nerves to leave through a common window. The kiwi's relatively large brain and small skull possibly account for a reversal of the paleognathous condition.
4. The chorda tympani is missing in all paleognathous species. The description of a vestigial chorda in the American rhea by Müller (1963) has been confirmed here. Its presence probably varies individually; I have found it in only one specimen on the left side. As discussed earlier, previous descriptions of a chorda in the ostrich and kiwi confuse it with the recurrent nerve. Since the chorda tympani is a shared character in vertebrates, its absence in paleognathous birds is a derived character.
5. The paratympanic organ, which is an ancestral vertebrate sensory organ and is present in neognathous birds, is missing in paleognathous birds. Because of outgroup relationships, the paleognathous condition is considered an apomorphy. However, reduction of a character is always a weak phylogenetic argument.

On the basis of this investigation of middle ear structures, only these five apomorphic complex characters can be recognized as characterizing paleognaths. They support earlier opinions (Bock 1963; Cracraft 1974; see 1.5.2) that paleognathous birds are monophyletic. However, in a phylogenetically strict sense, these middle ear characters are rather weak phylogenetic arguments. Two characters (the chorda tympani and the paratympanic organ) are simple reductions of the ancestral condition found in neognathous birds. One character (the cover of the pharyngotympanic tube) represents a gradual change of a structural trait whose intermediate neognathous forms could be overlooked. The exit of the ninth and tenth cranial nerve is not consistently found in paleognathous birds and its absence in the kiwi requires special consideration. The complex and specific form of the quadratocranial articulation and the obvious reduction of kinesis appear to be the most reliable characters.

It must be pointed out that the absence of a foramen of Huxley in paleognathous birds and its presence in neognathous hatchlings (!) are due to differences in developmental stages rather than phylogenetic differences. A foramen of Huxley is present in embryonic neognathous birds and young ostrich embryos. Its presence or absence therefore cannot be used as a phylogenetic character. In conclusion, five middle ear characters are recognized as supporting the monophyly of paleognathous birds. A critical evaluation of these characters shows that none is a really strong argument that could be used to establish the phylogeny without previous knowledge. However, they nicely support previous hypotheses about the monophyly of paleognathous birds. Nevertheless, these characters tell us nothing about the position of paleognaths within the Aves. All middle ear characters are derived in paleognaths, but neognathous birds as well as other sauropsids show the ancestral condition. Thus, this study cannot help resolve the question whether the palaeognathous birds are a sister group of neognathous birds (as suggested in Fig. 1) or whether they are a derived monotaxon within the neognathous birds. Middle ear characters cannot be used to test the phylogenetic relationships within Struthioniformes.

5 Summary

This study provides new anatomical data, morphometric measurements, and functional hypotheses about the middle ear of paleognathous and a basal neognathous bird. The introduction discusses current theoretical concepts and basic functional approaches in vertebrate morphology. A cautious and somewhat skeptical evaluation of the techniques used, errors made, and personal experience gained in this study forms the background of the interpretation of the structural data. The comparative discussion extracts phylogenetic information from the structural data.

The external ear opening of paleognathous species is large and unprotected. The auricular feathers show no structural specialization and provide only incomplete opercularization. No muscles insert into the external ear opening. Special attention has been paid to modifications in the cassowary's skin. The skin of its head has developed cavernous blood sinuses that may be flooded and thus inflate the head during booming display.

The external auditory meatus of paleognathous birds is large, bends ventrally, and reaches the tympanic membrane from the ventral side. No peculiarities can be described.

The osteology of the tympanic region is described in detail. The fusion of all bony elements and the ossification of connective tissue are important features of the avian middle ear region. The fusion of bone is a necessary prerequisite for extended pneumatization. The ossification of connective tissue, however, complicates morphological description since it deviates from morphologically defined bones. Especially in the neognathous button quail, most of the ventral wall of the middle ear cavity consists of ossified connective tissue; it also includes elements of the extracolumella.

The fixation of the eardrum to the bony wall of the tympanic cavity is described in detail. The fixation includes the kinetic quadrate and loose connective tissue in the ventro-lateral part of the middle ear cavity. Movements of the bill must change the tension of the eardrum and thus affect the hearing of birds.

The recesses of the middle ear cavity have been investigated using X-ray computed tomography. Using this technique, subsequent three-dimensional reconstruction provides unusual and unique insights into the anatomy of such "nonstructures" as air-filled cavities. It has been shown that three tympanic recesses are a character shared by all birds. Of more functional importance is the interaural pathway provided by the anterior tympanic recess and connecting the contralateral middle ear cavities. This structure is present in all recent and mesozoic birds. It has been suggested that the interaural pathway might function as a sound pressure gradient receptor. Physiological data are controversial but in many cases supportive. Pneumatic air volumes have been measured and preliminary resonance frequencies calculated. Low resonance frequencies might correlate with the low frequencies of the calls of most paleognathous birds.

The anatomy of the pharyngotympanic tube may be used in phylogenetic analysis since, in clear contrast to neognathous birds, its bony cover is very short.

The pneumatization of the mandible and the quadrate reveal no differences from neognathous birds. Pneumatization must be a shared character in all birds, if not sauropsids.

The quadratocranial arthrology shows structural peculiarities in paleognaths. The quadrate is monocondylic in all species. The articulation is a synchondrosis, which restricts wide excursions of the quadrate. In addition, ligaments firmly fix the otic process of the quadrate to the zygomatic process and the lateral wall of the basicranium. The zygomatic process lies very close to the quadrate, also restricting kinesis. This complex of characters can be found only in paleognathous birds and is functionally interpreted as a device to reduce active kinesis of the skull while retaining flexibility to counter external forces during jaw movements. The quadratocranial articulation is a derived character in paleognathous birds. The button quail represents the ancestral and neognathous condition.

Stapes and extracolumella are described in hatchlings and adults of all species. In paleognathous birds, their structure generally seems to be rather uniform. However, tinamous are unique, with a bifurcated infracolumellar process. Special attention is paid to the intracolumellar junction (between the bony stapes and cartilaginous extracolumella) and the fixation of the infracolumellar process to the wall of the tympanic cavity. Both features are considered important structures for a functional model of the middle ear. A hypothetical lever system is suggested, which would contribute an amplification function of ca. 1.6 to the transfer function of the middle ear.

Ligaments and muscles of the middle ear are described and discussed in terms of comparative morphology. The paratubal muscle was previously described only for the American rhea. It has now been found in most of the other species including turnix. Where not present, it is replaced by the inferior and superior drum-tubal ligament. The stapedial muscle is uniform in all birds.

The pathways of cranial and sympathetic nerves related to the middle ear cavity have been investigated. The chorda tympani is missing in paleognathous birds. This may be seen as a derived character for paleognaths. In all paleognathous birds except the kiwi, the ninth and tenth cranial nerves leave the braincase through a common window. This is peculiar since in all other birds, including the kiwi, the ninth cranial nerve leaves the braincase through the aquaeductus cerebri. However, the kiwi condition has tentatively been suggested to be secondarily neognathous.

The paratympanic organ is missing in paleognathous birds but present in the neognathous button quail.

The course of the stapedial artery and internal carotid shows no peculiarities that could be used for a phylogenetic analysis.

A detailed model of the functional morphology of the avian middle ear is suggested. Based on the calculation of the area ratio between the eardrum and clipeolus, a morphological lever system, and the curved membrane effect, the overall amplification function is estimated to be ca. 75. The morphological impedance could not be quantified, but aspects such as material components, mass reactance, and friction are discussed. Structural peculiarities such as the pneumatization of the stapes are discussed as representing structural adaptation to reduce mass reactance. Stiffness reactance is the only aspect of morphological impedance that can be controlled actively by the activity of middle ear muscles. This factor is probably an important aspect in the ear's adjustment of its dynamic range to actual sound pressure. However, considerable discrepancies between structural models and physiological input–output measurements show that these models do not correspond perfectly to the functional properties of the middle ear.

The structural data presented in this study are subjected to phylogenetic analysis. The short bony tunnel of the pharyngotympanic tube, the quadratocranial articulation, loss of the chorda tympani, the common exit of the ninth and tenth cranial nerves, and the loss of the paratympanic organ are shared derived characters in the paleognaths. Evidence supports the monophyly of paleognaths. The position of paleognathous birds within the birds cannot be elucidated using these characters because other sauropsids besides neognathous birds show the ancestral character state.

Acknowledgments. All airbrush drawings and most of the black-and-white drawings in this study were done by Margret Roser, Tübingen. I am deeply indebted to her for her effective collaboration, her unsurpassed skill in preparing the anatomical drawings, and her cooperation in an atmosphere of friendship. Her work may help clarify many details where words failed to describe structures appropriately. I would like to express my gratitude to Dietrich Starck, Hans-Rainer Duncker, and Walter J. Bock for the valuable material they made available for this study, for their technical support, and for their discussions during the preparation of this manuscript. I gratefully acknowledge the comments and suggestions of Barbara Helm, Walter J. Bock, and Dominique Homberger on an earlier version of this manuscript. Their suggestions were very helpful in improving the manuscript. I thank Tim and Lotte for growing along with the manuscript.

6 References

Anisimov VD, Iljitschew WD (1980a) The peripheral region of the penguin's sound-sensitive system. Morphol Aspekt Zbol (Mosc):130–147

Anisimov VD, Iljitschew WD (1980b) Sound-conducting functions of the middle ear in birds. I. Structural adaptations of specialized systems. Zool J (Mosc) 59:1706–1717

Ax P (1989) Homologie in der Biologie – ein Relationsbegriff im Vergleich von Arten. Zool Beitr NF 32:487–496

Baumel J, Gerchmann L (1968) The avian intercarotid anastomosis and its homologue in other vertebrates. Am J Anat 122:1–18

Baumel J, King AS, Breazile JE, Evans HE, Vanden Berge JC (1993) Handbook of avian anatomy: nomina anatomica avium, 2nd edn. Nuttall Ornithological Club, Cambridge, MA (Publications of the Nuttall Ornithological Club, vol 23)

Baumel J, King AS, Lucas AM, Breazile JE, Evans HE (1979) Nomina anatomica avium. Academic, London

Beehler BM, Pratt TK, Zimmermann DA (1986) Birds of New Guinea. Princeton University Press, Princeton (Handbook no 9 of the Wau Ecology Institute)

Blake ER (1977) Spheniscidae (penguins) to Laridae (gulls and allies). University of Chicago Press, Chicago (Manual of neotropical birds, vol 1)

Bock WJ (1963) The cranial evidence for ratite affinities. 13th International Ornithological Congress, Ithaca, pp 39–54

Bock WJ (1969) The origin and radiation of birds. Ann NY Acad Sci 167:147–155

Bock WJ (1977) Foundations and methods of classical evolutionary classification. NATO Adv Study Inst Ser A Life Sci 14:57–82

Bock WJ (1989) The homology concept: its philosophical foundation and practical methodology. Zool Beitr NF 32:327–353

Bock WJ (1990) From biologische anatomie to ecomorphology. Neth J Zool 40:245–277

Bock WJ (1991) Explanations in konstruktionsmorphologie and evolutionary morphology. In: Schmidt-Kittler N, Vogel K (eds) Constructional morphology and evolution. Springer, Berlin Heidelberg New York, pp 9–29

Bock WJ, Bühler P (1990) The evolution and biogeographical history of the palaeognathous birds. 100th International DO-G Meeting, Current Topics Avian Biol, Bonn 1988, pp 31–36

Bock WJ, McEvey A (1969) Osteology of *Pedionomus torquatus* (Aves: Pedionomidae) and its allies. Proc R Soc Vict 82:187–232

Bock WJ, von Wahlert G (1965) Adaptation and the form-function complex. Evolution 19:269-299

Bock WJ, Morony JJ, Farrand J (1975) Reference list of the birds of the world. American Museum of Natural History, New York

Borg E, Counter SA, Rydquist B (1979) Contraction properties and functional morphology of the avian stapedius muscle. Acta Otolaryngol (Stockh) 88:20–26

Brock GT (1937) The morphology of the ostrich chondrocranium. Proc Zool Soc Lond Ser B 107:225–243

Brown L, Urban EK, Newman K (1983) The birds of Africa, vol 1. Academic, New York

Bubien-Waluszewska A (1981) The cranial nerves. In: King AS, McLelland J (eds) Form and function in birds, vol 2. Academic, New York

Bühler P (1981) Functional anatomy of the avian jaw apparatus. In: King AS, McLelland J (eds) Form and function in birds, vol 2. Academic, New York

Bühler P (1987) Zur Strategie des Beutefangs der Nachtschwalben (Caprimulgidae). J Ornithol 128:488–491

Chatterjee S (1991) Cranial anatomy and relationship of a new triassic bird from Texas. Philos Trans R Soc Lond [Biol] 332:277–342

Coates BJ (1985) Nonpasseriformes. Dove, Alderley (Birds of Papua Neu Guinea, vol 1)

Cohen YE, Hernandez HN, Saunders JC (1992a) Middle ear development. II. Morphometric changes in the conducting apparatus of the chick. J Morphol 212:257–267

Cohen YE, Rubin DM, Saunders JC (1992b) Middle ear development. I. Extra-stapedius response in the neonatal chicken. Hear Res 58:1–8

Coles RB, Lewis DB, Hill KG, Hutchings ME, Gower DM (1980) Directional hearing in the Japanese quail *Coturnix coturnix japonica*. II. Cochlear physiology. J Exp Biol 86: 153–170

Cords E (1904) Beiträge zur Lehre vom Kopfnervensystem der Vögel. Bergmann, Wiesbaden (Medical dissertation, University of Freiburg)

Counter SA (1978) Ultrastructural features of neuromuscular junctions in the stapedius muscle of *Gallus gallus*. J Morphol 158:361–366

Counter SA, Borg E (1979) Physiological activation of the stapedius muscle in *Gallus gallus*. Acta Otolaryngol (Stockh) 403:13–19

Cracraft J (1974) Phylogeny and evolution of ratite birds. Ibis 116:494–521

Cracraft J (1981) Toward a phylogenetic classification of the recent birds of the world. Auk 98:681–714

Cracraft J (1986) The origin and early diversification of birds. Palaeobiology 12:383–399

Cracraft J (1988) The major clades of birds. In: Benton MJ (ed) The phylogeny and classification of the tetrapodes. Clarendon, London (The systematic association special volume 35A/1)

Cracraft J, Mindel DP (1989) The early history of modern birds: a comparison of molecular and morphological evidence. In: Fernholm B, Bremer K, Jörnvall H (eds) The hierarchy of life. Molecules and morphology in phylogenetic analyses. Elsevier, Amsterdam (Nobel symposium, vol 70)

Cramp S, Simmons KEL (1977). Ostrich – Ducks. Oxford University Press, New York (Handbook of the birds of Europe, the Middle East and North Africa, vol 1)

Crompton AW (1953) The development of the chondrocranium of *Spheniscus demersus* with special reference to the columella auris of birds. Acta Zool (Stockh) 34:71–146

Currie PJ (1985) Cranial anatomy of *Stenonychosaurus inequalis* (Saurischia, Theropoda) and its bearing on the origin of birds. Can J Earth Sci 22:1643–1658

Cuvier G (1798–1805) Leçons d'anatomie comparée, 5 vols., Paris

De Beer GR (1956) The evolution of ratites. Bull Br Mus (Nat Hist) Zool 4:60–70

De Beer GR, Barrington BA (1934) The segmentation and chondrification of the skull of the duck. Philos Trans R Soc Lond [Biol] 223:411–467

De Kock JM (1955) The cranial morphology of *Sturnis vulgaris vulgaris* Linneus. Ann Univ Stellenbosch Ser A 31:153–177

De Kock JM (1987) The development of the chondrocranium of *Melopsittacus undulatus*. Adv Anat Embryol Cell Biol 104:1–70

Dombrowsky B (1925) Das Mittelohr der Vögel. Rev Zool Russe 5:15–35

Dullemeijer P (1974) Concepts and approaches in animal morphology. Van Gorcum, Assen

Edgeworth FH (1935) The cranial muscles of vertebrates. Cambridge University Press, Cambridge

Elzanowski A (1989) Ontogeny and evolution of ratites. Intern Congr Ornithol 19:2037–2046

Elzanowski A (1995) Mesozoic birds and avian phylogeny. Cour Forschungsinst Senckenb 181:481ff

Engelbrecht DZ (1958) The development of the chondrocranium of *Pyromelana orix orix*. Acta Zool (Stockh) 39:115–199

Feduccia A (1974) Morphology of the bony stapes in New and Old World suboscines: new evidence for common ancestry. Auk 91:427–429

Feduccia A (1975a) Morphology of the bony stapes (columella) in the passeriformes and related groups: evolutionary implications. Univ Kansas Mus Nat Hist Misc Publ 63

Feduccia A (1975b) The bony stapes in the Upupidae and Phoeniculidae: new evidence for common ancestry. Wilson Bull 87:416–417

Feduccia A (1976) Morphology of the bony stapes in *Philepitta* and *Neodrepanis*: new evidence for suboscine affinities. Auk 93:169–170

Feduccia A (1977) A model for the evolution of perching birds. Syst Zool 26:19–31

Feduccia A (1978) Morphology of the bony stapes (Columella) in birds: evolutionary implications. 17th International Congress of Ornithology, Berlin, pp 620–624

Feduccia A (1980) The age of birds. Harvard University Press, Cambridge MA

Folch A (1992) Struthioniformes. In: del Hoyo J, Elliott A, Sargatal J (eds) Ostrich to ducks. Lynx, Barcelona (Handbook of the birds of the world, vol 1)

Fourie S (1955) A contribution to the cranial morphology of *Nyctisyrigmus pectoralis pectoralis* with special reference to the palate and cranial kinesis. Ann Univ Stellenbosch Ser A 31:179–215

Frank GH (1954) The development of the chondrocranium in the ostrich. Ann Univ Stellenbosch Ser A 30:197–248

Frank GH, Smit AL (1974) The early ontogeny of the columella auris of Crocodilus niloticus and its bearing on problems concerning the upper end of the reptilian hyoid arch. Zool Afr 9:59–88

Frank GH, Smit AL (1976) The morphogenesis of the avian columella auris with special reference to *Struthio camelus*. Zool Afr 11:159–182

Freye HA (1952/1953) Das Gehörorgan der Vögel. Wiss Z Martin Luther Univ Halle Wittemberg Math Naturwiss Reihe 2(5):267–297

Freye-Zumpfe H (1952/1953) Befunde im Mittelohr der Vögel. Wiss Z Martin Luther Univ Halle Wittemberg Math Naturwiss Reihe 2:445–461

Gadow H (1891) Vögel. In: Bronn HG (ed) Klassen und Ordnungen des Thierreichs, vol 6, sect 4. C.F. Winter, Leipzig

Gaudin EP (1968) On the middle ear of birds. Acta Otolaryngol (Stockh) 65:316–326

Gaupp E (1898) Ontogenese und Phylogenese des schalleitenden Apparates bei den Wirbeltieren. Ergeb Anat Entwicklungsgesch 8:990–1149

Gaupp E (1913) Die Reichertsche Theorie (Hammer-, Amboß- und Kieferfrage). Arch Anat Physiol Abt Anat Suppl, pp 1–416

Gauthier J, Padian K (1985) Phylogenetic, functional and aerodynamic analyses of the origin of birds and their flight. In: The beginnings of birds. Proceedings of the International Archaeopteryx Conference, Eichstätt 1984, pp 185–197

Glutz von Blotzheim U (1958) Zur Morphologie und Ontogenese von Schultergürtel, Sternum und Becken von *Struthio*, *Rhea* und *Dromiceius*. Rev Suisse Zool 65:609–772

Goldschmid A (1972a) Die Entwicklung des Craniums der Mausvögel (Coliidae, Coliiformes, Aves). I. Die Frühentwicklung des Chondrocraniums. Gegenbaurs Morph Jahrb Leipzig 118:105–138

Goldschmid A (1972b) Die Entwicklung des Craniums der Mausvögel (Coliidae, Coliiformes, Aves): II. Die Entwicklung des Chondrocraniums bis zum Auftreten der dermalen Verknöcherungen. Gegenbaurs Morphol Jahrb 118:274–304

Goldschmid A (1972c) Die Entwicklung des Craniums der Mausvögel (Coliidae, Coliiformes, Aves). III. Ausgestaltung und Umbau des Chondrocraniums bis zum Auftreten der Ersatzverknöcherungen. Gegenbaurs Morphol Jahrb 118:369–413

Golubeva TB (1972) The reflex activity of the tympanal muscle in the owl *Asio otus*. Zh Evol Biol Fiziol 8:173–181

Gummer AW, Smolders JWT, Klinke R (1989a) Mechanics of a single-ossicle ear. I. The extra-stapedius of the pigeon. Hear Res 39:1–14

Gummer AW, Smolders JWT, Klinke R (1989b) Mechanics of a single-ossicle ear. II. The columella footplate of the pigeon. Hear Res 39:15–26

Haubitz B, Prokop M, Döhring W, Ostrom JH, Wellnhofer P (1988) Computed tomography of *Archaeopteryx*. Palaeobiology 14:206–213

Haubitz B, Wellnhofer P (1989) Computertomographie bei *Archaeopteryx*: weiteres Vogelmerkmal. Spektrum Wiss 20:23–30

Harvey PH, Pagel MD (1991) The comparative method in evolutionary biology. Oxford University Press, Oxford (Oxford Series in Ecology and Evolution)

Hennig W (1950) Grundzüge einer Theorie der phylogenetischen Systematik. Deutscher Zentralverlag, Berlin

Hill KG, Lewis DB, Hutchings ME, Coles RB (1980) Directional hearing in the Japanese quail (*Coturnix coturnix japonica*). I. Acoustic properties of the auditory system. J Exp Biol 86:135–151

Hofer H (1948) Untersuchungen über den Bau des Vogelschädels, besonders über den der Spechte und Steißhühner. Zool Jahrb Abt Anat 69:1–158

Hoffmann CK (1889) Over de ontwikkelingsgeschiedenis van het gehoororgaan en de morphologische beteekenis van het gehoorbeentje bij de reptilien. Natuurkol Verh Konigkl Acad 28

Holmgren N (1955) Studies on the phylogeny of birds. Acta Zool (Stockh) 36:243–328

Homberger DG (1986) The lingual apparatus of the African grey parrot, *Psittacus erithacus* Linné (Aves: Psittacidae): description and theoretical mechanical analysis. Ornithol Monogr 39:1–233

Homberger DG (1988) Models and tests in functional morphology: the significance of description and integration. Am Zool 28:217–229

Houde P (1988) Palaeognathous birds from the early tertiary of the northern hemisphere. Nuttall Ornithological Club, Cambridge, MA (Publications of the Nuttall Ornithological Club, vol 22)

Hsieh TM (1951) The sympathetic and parasympathetic nervous system of the fowl. PhD thesis, University of Edinburgh

Huxley TH (1867) On the classification of birds; and on the taxonomic value of the modifications of certain cranial bones observable in the class. Proc Zool Soc Lond 27:415–472

Iljitschew WD (1967) Die akustische Lokalisation der Vögel (in Russian). Akademijy nauk SSSR, Moscow

Iljitschew WD (1968) Adaptationen des Gehörsystems der Vögel und ihre Rolle in der Evolution (in Russian) Zurnal Obshch Biol 29:32–46

Iljitschew WD (1974) Adaptationsökologische Parallelismen – mosaikartige Evolution. Das Hörsystem der Vögel als Objekt der funktionellen Morphologie. Biol Zentralbl 93: 165–180

Jollie MT (1957) The head skeleton of the chicken and remarks on the anatomy of this region in other birds. J Morphol 100:389–436

Kartaschew NN, Iljitschew WD (1964) Über das Gehörorgan der Alkenvögel. J Ornithol 105:113–136

Knudsen EI (1980) Sound localization in birds. In: Popper AN, Fay RR (eds) Comparative studies of hearing in vertebrates. Springer, Berlin Heidelberg New York

Knudsen EI, Konishi M (1979) Mechanisms of sound localization in the barn owl (*Tyto alba*). J Comp Physiol 133:13–21

Krause G (1901) Die Columella der Vögel (Columella auris avium). Ihr Bau und dessen Einfluß auf die Feinhörigkeit. Neue Untersuchungen und Beiträge zur comparativen Anatomie des Gehörorganes. Friedländer, Berlin

Kesteven HL (1942) The ossification of the avian chondrocranium with special reference to that of the emu. Proc Linn Soc New South Wales 67:213–237

Kühne R, Lewis B (1985) External and middle ears. In: King AS, McLelland J (eds) Form and function in birds, vol 3. Academic, London

Lang C (1954) Beiträge zur Entwicklungsgeschichte des Kopfskelettes von *Melopsittacus undulatus*. Morphol Jahrb 94:335–390

Lang C (1956) Das Cranium der Ratiten mit besonderer Berücksichtigung von *Struthio camelus*. Z Wiss Zool 159:165–224

Leisler B, Winkler H (1985) Ecomorphology. Curr Ornithol 2:155–186

Leisler B, Winkler H (1991) Ergebnisse und Konzepte ökomorphologischer Untersuchungen an Vögeln. J Ornithol 132:373–425

Lewald J (1990) The directionality of the ear of the pigeon (*Columba livia*). J Comp Physiol [A] 167:533–543

Lorenz K (1936) Über eine eigentümliche Verbindung branchialer Hirnnerven bei *Cypselus apus*. Morphol Jahrb 77:305–325

Lowe PR (1926) Notes on the quadrate as a factor in avian classification. Ibis 2:152–189

Lowe PR (1928) Studies and observations bearing on the phylogeny of the ostrich and its allies. Proc Zool Soc Lond 48:185–247

Lowe PR (1944) Some additional remarks on the phylogeny of struthiones. Ibis 86:37–42
Lucas AM, Stettenheim PR (1972) Avian anatomy, integument, 2 vols. Agricultural Research Service, United States Department of Agriculture, Michigan Agricultural Experiment Station, Washington, DC (Agriculture Handbook, vol 362)
Lutz H (1942) Beiträge zur Stammesgeschichte der Ratiten. Vergleich zwischen Emu-Embryo und entsprechenden Carinatenstadien. Rev Suisse Zool 49:299–399
Macdonald JD (1973) Birds of Australia. Witherby, London
Macke T (1969) Die Entwicklung des Craniums von *Fulica atra* L. Gegenbaurs Morphol Jahrb 113:229–294
Manley GA (1972) Frequency responses of the middle ear of the tokay gecko. J Exp Zool 181:159–168
Manley GA (1990) Peripheral hearing mechanisms in reptiles and birds. Springer, Berlin Heidelberg New York (Zoophysiology, vol 26)
May W (1961) Die Morphologie des Chondrocraniums und Osteocraniums eines Waldkauzembryos (*Strix aluco* L.). Z Wiss Zool 166:134–202
Mayr E (1969) Principles of systematic zoology, 1st edn. McGraw-Hill, New York
Mayr E (1974) Cladistic analysis or cladistic classification? Z Zool Syst Evolutionsforsch 12:94–128
Mayr E (1990) Die drei Schulen der Systematik. Verh Dtsch Zool Ges 83:263–276
Mayr E, Amadon D (1951) A classification of recent birds. Am Mus Novit 1496:2–42
McDowell S (1948) The bony palate of birds. I. The palaeognathae. Auk 65:520–549
Meise W (1963) Verhalten der straußartigen Vögel und Monophylie der Ratitae. 13th International Ornithological Congress, pp 115–125
Moiseff A, Konishi M (1981) The owl's interaural pathway is not involved in sound localization. J Comp Physiol 144:299–304
Møller AR (1965) An experimental study of the acoustic impedance of the middle ear and its transmission properties. Acta Otolaryngol (Stockh) 60:129–149
Müller HJ (1961) Die Morphologie und Entwicklung des Craniums von *Rhea americana* Linné. I. Das knorpelige Neurocranium. Z Wiss Zool 165:221–319
Müller HJ (1963) Die Morphologie und Entwicklung des Craniums von *Rhea americana* Linné. II. Viszeralskelett, Mittelohr und Osteocranium. Z Wiss Zool 168:35–118
Murie J (1867) On the tracheal pouch of the emu (*Dromæus novæ-hollandiæ*, Vieill). Proc Zool Soc Lond 34:405–415
Nickel R, Schummer A, Seiferle E, Siller WG, Wight PAL (1977) Anatomy of the domestic birds. Parey, Berlin Springer, Berlin, Heidelberg New York
Norberg RÅ (1977) Occurrence and independent evolution of bilateral ear asymmetry in owls and implications on owl taxonomy. Philos Trans R Soc Lond [Biol] 280:375–408
Norberg RÅ (1977) Skull asymmetry, ear structure and function, and auditory localization in Tengmalm's owl, *Aegolius funereus* (Linné). Philos Trans R Soc Lond [Biol] 282:325–410
Oeckinghaus H (1978) Einfluß von Kontraktionen des Mittelohrmuskels auf die Entladungen von Hörnervenelementen beim Star. Verh Dtsch Zool Ges 7:191
Oeckinghaus H, Schwartzkopff J (1983) Electrical and acoustical activation of the middle ear muscle in a songbird. J Comp Physiol [A] 150:61–67
Oldenstam RA (1925) Over het z.g.n. vliegeorgaan van Vitali in het middenoor der vogels. Dissertation, University of Groningen
Olson SL (1985) The fossil record of birds. In: Farner DS, King JR, Parkes KC (eds) Avian biology, vol 8. Academic, New York
Osche G (1973) Das Homologisieren als eine grundlegende Methode der Phylogenetik. Aufsätze Reden Senckenb Naturf Ges 24:155–165
Parker TJ (1888) Preliminary note on the anatomy and development of Apteryx. Proc R Soc Lond 43:391–397
Parker TJ (1891) Observations on the anatomy and development of Apteryx. Philos Trans R Soc Lond [Biol] 182:25–134
Parker WK (1866a) On the structure and development of the skull in the ostrich tribe. Philos Trans R Soc Lond [Biol] 156:113–183
Parker WK (1866b) On the osteology of gallinaceous birds and tinamous. Trans Zool Soc Lond Ser B 5:149–241

Parker WK (1869) On the structure and development of the skull of the common fowl (*Gallus domesticus*). Philos Trans R Soc Lond [Biol] 159:755–807

Parker WK (1875) On the morphology of the skull in the woodpeckers (Picidae) and wrynecks (Jyngidae). Trans Linn Soc Lond 1:1–22

Parkes KC, Clark GA Jr (1966) An additional character linking ratites and tinamous, and an interpretation of their monophyly. Condor 68:459–471

Payne RS (1971) Acoustic location of prey by barn owls (*Tyto alba*). J Exp Biol 54:535–573

Peters DS (1987) Mechanische Unterschiede paläognather und neognather Vogelschädel. Nat Mus 117:173–182

Plate L (1918) Über Drüsen und Lymphknoten in der Ohrfalte der Truthenne und des Auerhahns. Arch Mikrosk Anat 91:208–217

Pohlmann AG (1921) The position and functional interpretation of the elastic ligaments in the middle-ear region of Gallus. J Morphol 35:229–262

Popper KL (1969) Logik der Forschung, 3rd edn. Mohr, Tübingen

Pumphrey RJ (1961) Sensory organs: hearing. In: Marshall AJ (ed) Biology and comparative physiology of birds. Academic, New York

Pycraft WP (1901) Some points in the morphology of the palate of the neognathae. J Linn Soc Lond 28:343–357

Reichert C (1837) Über die Visceralbogen der Wirbeltiere im Allgemeinen und deren Metamorphosen bei den Vögeln und Säugetieren. Arch Anat Physiol Wiss Med, pp 120–222

Remane A (1952) Die Grundlagen des natürlichen Systems, der vergleichenden Anatomie und der Phylogenetik. Geest und Portig, Leipzig

Remane J (1989) Die Entwicklung des Homologiebegriffs seit Adolf Remane. Zool Beitr NF 32:497–504

Romeis B (1989) Mikroskopische Technik, 17th edn. Urban und Schwarzenberg, Munich

Rosowski JJ, Saunders JC (1980) Sound transmission through the avian interaural pathway. J Comp Physiol [A] 136:183–190

Saayman JE (1963) In Bydrae tot die Skedelmorfologie van *Passer m. melanurus*. Ann Univ Stellenbosch 38A:189–213

Saiff EI (1974) The middle ear of the skull of birds. The procellariiformes. Zool J Linn Soc 54:213–240

Saiff EI (1976) Anatomy of the middle ear region of the avian skull: sphenisciformes. Auk 93:749–759

Saiff EI (1978a) Middle ear anatomy of the struthioniformes. 17th International Congress on Ornithology, Berlin, pp 631–634

Saiff EI (1978b) The middle ear of the skull of birds: the pelecaniformes and ciconiiformes. Zool J Linn Soc 63:315–370

Saiff EI (1981) The middle ear of the skull of birds: the ostrich: *Struthio camelus* L. Zool J Linn Soc 73:201–212

Saiff EI (1982) The middle ear of the skull of the kiwi. Emu 82:75–79

Saiff EI (1983) The anatomy of the middle ear region of the rheas (Aves: Rheiformes, Rheidae). Hist Nat Corrientes Argentina 3(6):45–55

Saiff EI (1988) The anatomy of the middle ear of the tinamiformes (Aves: Tinamidae). J Morphol 196:107–116

Saunders JC (1985) Auditory structure and function in the bird middle ear: an evaluation by SEM and capacitive probe. Hear Res 18:253–268

Saunders JC, Johnstone BM (1972) A comparative analysis of middle ear function in non-mammalian vertebrates. Acta Otolaryngol (Stockh) 73:353–361

Sauer EGF (1970) Interspecific behavior of the South African ostrich. Ostrich 8 Suppl: 91–103

Schestakowa GS (1934) Zur Frage über die Homologie der Gehörknöchelchen. Die Entwicklung des schalleitenden Apparates der Vögel. Bull Soc Nat Mosc S Biol 43:225–237

Schoonees J (1963) Some aspects of the cranial morphology of *Colius indicus*. Ann Univ Stellenbosch 38A:215–246

Schwalbe G (1890) Über den Gehörgangwulst der Vögel. Arch Anat Physiol (Anat Abt): 42–63

Schwartzkopff J (1950) Beitrag zum Problem des Richtungshörens bei Vögeln. Z Vergl Physiol 32:319–327
Schwartzkopff J (1952a) Über den Gehörsinn der Vögel. J Ornithol 93:91–103
Schwartzkopff J (1952b) Untersuchungen über die Arbeitsweise des Mittelohres und das Richtungshören der Singvögel unter Verwendung von Chochlea-Potentialen. Z Vergl Physiol 34:46–68
Schwartzkopff J (1955) On the hearing of birds. Auk 72:340–347
Schwartzkopff J (1957) Die Größenverhältnisse von Trommelfell, Columella-Fußplatte und Schnecke bei Vögeln verschiedenen Gewichts. Z Morphol Okol Tiere 45:365–378
Schwartzkopff J (1973) Mechanoreception. In: Farner DS, King JR, Parkes KC (eds) Avian biology, vol 3. Academic, New York
Sibley CG, Ahlquist JE (1990) Phylogeny and classification of birds. A study in molecular evolution. Yale University Press, New Haven
Sibley CG, Monroe BL Jr (1990) Distribution and taxonomy of birds of the world. Yale University Press, New Haven
Sibley CG, Ahlquist JE, Monroe BL (1988) A classification of the living birds of the world based on DNA-DNA hybridization studies. Auk 105:409–423
Simonetta AM (1960) On the mechanical implications of the avian skull and their bearing on the evolution and classification of birds. Q Rev Biol 35:206–220
Skutch A (1983) Birds of tropical America. Texas University Press, Austin (Corrie Herring Hooks series, no 5)
Slaby O (1951) Le développement du chondrocrane du cormorant (*Phalacrocorax carbo* L.) au point de vue de l'évolution. Bull Int Acad Tchèque Sci 9:1–47
Smit AL, Frank GH (1979) Aspects of the ontogenesis of the avian columella auris. S Afr Tydskr Dierkd 14:23–35
Smith G (1905) The middle ear and columella of birds. Q J Microsc Sci 48:11–22
Sokal RR, Sneath PHA (1963) Principles of numerical taxonomy. Freeman, San Francisco
Sonies F (1907) Über die Entwicklung des Chondrocraniums bei den Vögeln. Petrus Camper 4:395–623
Starck D (1955) Die endokraniale Morphologie der Ratiten, besonders der Apterygidae und Dinornithidae. Morphol Jahrb 96:14–72
Starck D (1982) Vergleichende Anatomie der Wirbeltiere auf evolutionsbiologischer Grundlage, vol 3. Springer, Berlin Heidelberg New York
Starck JM (1989) Zeitmuster der Ontogenesen bei nestflüchtenden und nesthockenden Vögeln. Cour Forschungsinst Senckenb 114:1–319
Starck JM (1991a) Biogeography and life history of *Turnix suscitator* Gmelin 1789. Small adult body size as a consequence of selection for rapid growth. Z Zool Syst Evolutionsforsch 29:213–237
Starck JM (1991b) Zur Biologie des Binden-Laufhühnchens *Turnix suscitator* Gmelin 1789. Trop Vogel 12:46–54
Starck JM (1992) Towards a model of the functional morphology of the avian middle ear. The transfer function. Zool Jahrb Anat 122:287–302
Starck JM (1993) Evolution of avian ontogenies. Curr Ornithol 10:275–366
Starck JM, Kriete A (1989) Morphometry and 3D-reconstruction of serial sections, methodological aspects. Eur J Cell Biol 48 Suppl 25:99–102
Storer RW (1971) Classification of birds. In: Farner DS, King JR, Parkes KC (eds) Avian biology, vol 1. Academic, New York
Stresemann E (1934) Sauropsida: Aves. de Gruyter, Berlin (Handbuch der Zoologie 7/2)
Suschkin P (1899) Zur Morphologie des Vogelskelettes. Nouv Mem Soc Imp Nat Mosc 16:1–163
Suschkin P (1905) Zur Morphologie des Vogelskelettes. Vergleichende Osteologie der normalen Tagraubvögel und die Fragen der Klassifikation. Nouv Mem Soc Imp Nat Mosc, pp 1–247
Swart PJ (1946) Die ontogenese van die "Desmognate" monddaktipe by Anas. Ann Univ Stellenbosch 24:1–20
Tiedemann F (1810) Anatomie und Naturgeschichte der Vögel. Landshut, Heidelberg (Zoologie, vol 2)

Toerien JM (1971) The developmental morphology of the chondrocranium of *Podiceps cristatus*. Ann Univ Stellenbosch Ser A 46:1–128

Tonkoff W (1900) Zur Entwicklungsgeschichte des Hühner-Schädels. Anat Anz 18:296–304

Tonndorf J, Khanna SM (1970) The role of the tympanic membrane in middle ear transmission. Ann Otol Rhinol Laryngol 79:743–754

Verheyen R (1960a) Les kiwis (Apterygiformes) dans les systèmes de classification. Bull Soc R Zool Anvers 15:3–11

Verheyen R (1960b) Contribution à l'ostéologie et à la systématique des Ratitae. Bull Soc R Zool Anvers 17:3–16

Versluys J (1904) Entwicklung der Columella auris bei den Lacertilien. Ein Beitrag zur Kenntnis der schalleitenden Apparate und des Zungenbeinbogens bei den Sauropsiden. Zool Jahrb Abt Anat Ontog 19:107–188

Von Bartheld CS (1990) Development and innervation of the paratympanic organ (Vitali organ) in chick embryos. Brain Behav Evol 35:1–15

Von Bartheld CS, Rubel EW (1989) The paratympanic organ: morphology, innervation and function. Soc Neurosci Abstr 15:375

Von Helmholtz H (1868) Die Mechanik der Gehörknöchelchen und des Trommelfells. Pflugers Arch 1:1–60

Von Békésy G (1960) Experiments in hearing. McGraw-Hill, New York

Von Tyne, Berger AJ (1976) Fundamentals in Ornithology. Wiley, New York

Vorster W (1989) The development of the chondrocranium of *Gallus gallus*. Adv Anat Embryol Cell Biol 113:1–77

Wada Y (1923) Beiträge zur vergleichenden Physiologie des Gehörgangs. Pflugers Arch 202:46–69

Webb M (1957) The ontogeny of the cranial bones, cranial peripheral and cranial parasympathetic nerves, together with a study of the visceral muscles of *Struthio*. Acta Zool (Stockh) 38:83–203

Weber E (1990) Zur Kraniogenese bei der Lachmöve (*Larus ridibundus* L.), zugleich ein Beitrag zur Rekonstruktion des Grundplanes der Vögel. Gegenbaurs Morphol Jahrb 136:335–387

Werner CF (1958) Der Canaliculus (Aquaeductus) cochleae und seine Beziehung zu den Kanälen der IX. and X. Hirnnerven bei den Vögeln. Zool Jahrb Abt Anat 77:1–8

Wetmore A (1981) The birds of the Republic of Panama. I. Tinamidae to Rynchopidae. Smithonian Institution, Washington (Smithonian miscellaneaous collections, vol 150/1)

Witmer LM (1990) The craniofacial air sac system of mesozoic birds (Aves). Zool J Linn Soc 100:327–378

Wolters HE (1975–1982) Die Vogelarten der Erde. Parey, Hamburg

Wong CJH, To EC, Schwarz DWF (1992) Location of motoneurons innervating the middle ear muscle of the chicken (*Gallus domesticus*). Hear Res 61:31–34

Zilles K, Schleicher A, Pehlemann FW (1982) How many sections must be measured in order to reconstruct the volume of a structure using serial sections. Microsc Acta 86:339–346

Zusi RL (1984) A functional and evolutionary analysis of rhynchokinesis in birds. Smithson Contrib Zool 395:1–40

7 Subject Index

8 List of Abbreviations and Synonyms Used in Avian Middle Ear Studies

Acc	Arteria carotis cerebralis	
Aci	Arteria carotis interna	Internal carotid
Aft	Pteryla auricularis	Anterior and posterior auricular feather tract
Amr	Apertura medialis recessus scalae tympani	Aquaeductus cochleae Canaliculus cochleae (Werner 1958) Apertura lateralis recessus scalae tympani Glossopharyngeal canal (Bubien-Waluszewska 1981)
Ap	Arteria palatinalis	
Asp	Arteria sphenopalatinalis	
Ast	Arteria stapedialis	Stapedial artery Arteria ophthalmica interna
ate	Ala tympanica ossis exoccipitalis	Tympanic wing of exoccipital bone
Ba	Basioccipitale	Os basioccipitale
Bsc	Basisphenoid complex	Basitemporal plate
CC	Canalis caroticus	Parabasal canal
Cl	Clipeolus	Clipeolus ossis columellae (Krause 1901) Basis columellae (NAA)
Clmq	Condylus mandibularis lateralis quadrati	
cm	Tympanic wing of exoccipitale	Ossification of metotic cartilage (from exoccipital bone)
Cm	Cartilago metotica	Metotic cartilage (see tympanic wing of exoccipital)
Cmmq	Condylus mandibularis medialis quadrati	
CMk	Cartilago Meckeli	Meckel's cartilage
CnVIIg	Connexus ggl. cervicalis cum facialis	
CnX	Connexus vago-glossopharyngealis	

Co	Columella (NNA)	Columella auis (Krause 1901) Os tympanicum (Parker 1876; cit. in Freye 1952/1953)
Cp	Crista parotica	
Cqo	Cotyla quadratica otici (NAA2)	
Cqs	Cotyla quadratica squamosi (NAA2)	
Ct	Chorda tympani	Ramus postrematicus nervi facialis
E	Extracolumella	Extracolumella (Parker 1876) Cartilago extracolumellaris (NAA) Pars extrastapedialis columellae
em	Extracellular matrix	
Fas	Fascia superficialis	Superficial fascia (Homberger 1986)
FH	Foramen Huxley	
fib	Fibroelastic margin of tympanic membrane	Margo fibroelasticus (NAA2)
Fj	Foramen jugulare	Vagal canal (Bubien-Waluszewska 1981)
FnX	Foramen nervi vagi	
Fo	Fenestra ovalis	
Fp	Foramen pseudorotundum	
Fpl	Foramen perilymphaticum	Oval window
Fr	Fenestra rotunda	
Fv	Foramen vestibulare	Vestibular window
Gcc	Ganglion craniale cervicale	Ggl. craniale cervicale Supracervical ganglion Ggl. supracervicale
GdIX	Ganglion distale n. glossopharyngei (Bubien-Waluszewska 1981)	Petrosal ganglion (Nickel et al. 1977)
Gla	Glandulae auriculares	Skin glands of auditory meatus
Gll	Glandulae lacrimales	
GVII	Ganglion geniculi	Facial ganglion Ggl. nervi facialis
Ifb	Promontorium	Interfenestral bar
In	Intracolumellar hinge Stipes extracolumellar junction	Articulatio intracolumellaris
Ip	Interaural pathway	
La	Ligamentum ascendens (Freye 1952/1953)	Ascending ligament Middle drum-tubal ligament (Pohlmann 1921)

		Extracolumellar ligament (Smith 1905)
Lan	Ligamentum annulare	Annular ligament
Lb	Lamina basalis	
Lcq	Ligamentum columello-squamosum (Pohlmann 1921; NAA)	Platner's ligament Ligamentum columello-quadratum Ligamentum Platneri
Lj	Ligamentum jugo-mandibulare	
lJ	Lower jaw	
lps	Lateral wing of parasphenoidal bone	Ala parasphenoidalis
Mae	Meatus acusticus externus	External auditory meatus External acustic meatus
mc	Cavitas metotica	Metotic cavity
Mcc	Musculus constrictor colli (NAA2)	Constrictor collis muscle
Mcci	Musculus constrictor colli intermandibularis (NAA2)	Intermandibular part of constricor collis muscle
Mdm	Musculus depressor mandibulae	Depressor mandible muscle
Me	Middle ear cavity	Cavum tympani (Freye 1952/1953) Cavitas tympanica (NAA) Middle ear sac Tympanic fossa (Saiff 1974)
mf	Microfibrillar elements	
Mp	Musculus paratubalis	Extrastapedial ligament (Smith 1905) Inferior and superior drum-tubal ligament (Pohlmann 1921)
Mst	Musculus stapedialis (Plate 1924)	Musculus columellae (NAA2) Stapedial muscle Musculus tensor tympani (Stresemann 1934; Freye 1952/1953 Pohlmann 1921) Musculus laxator tympani (in Starck 1982) M. occipito-tympanicus (Edgeworth 1935) M. extracolumellaris (Versluys 1904)
nV	Nervus trigeminus	Fifth cranial nerve
nV_1	Ramus ophthalmicus nervi V	
nV_2	Ramus maxillaris nervi trigemini	Maxillary nerve

nVII	Nervus facialis	Seventh cranial nerve
$nVII_1$	Nervus palatinalis	Ramus palatinalis nervi facialis Nervus petrosus major (Nickel et al. 1977)
$nVII_2$	Nervus hyomandibularis	Ramus hyomandibularis nervi facialis Nervus hyomandibularis (NAA)
nIx	Nervus glossopharyngeus	Ninth cranial nerve (Glossopharyngeal nerve)
nX	Nervus vagus	Tenth cranial nerve
nXII	Nervus hypoglossus	12th cranial nerve
$nrVII_2$	Nervus recurrens hyomandibularis	Connexus N. VII hyomandibularis cum $N.V_2$ Rami communicantes cum ramus secundus nervi trigemini (Cords 1904) Lesser superficial petrosal nerve (Hsieh 1951; Bubien-Waluszewska 1981) Communicationes petrosomaxillares et petrosomandibulares (Nickel et al. 1977)
nsp	Nervus sphenopalatinalis (Müller 1963)	Nervus petrosus major (Nickel et al. 1977) N. caroticus cerebralis (NAA)
nVid	Nervus vidianus	Nervus canalis pterygoidei (Nickel et al. 1977; Bubien-Waluszewska 1981)
Oa	Os articulare	Articular bone
Obo	Os basioccipitale	Basioccipital bone
Obp	Os basiparasphenoidale	Basiparasphenoid Os basisphenoidale
Oe	Os exoccipitale	Exoccipital bone
Opo	Os prooticum	Prootic bone
Opp	Os parasphenoidale	Parasphenoid
Oq	Os quadrati	Quadratum
Oso	Os supraoccipitale	Supraoccipital bone
Pa	Pennae auricularis	Pennulum
Pbp	Processus basipterygoideus	Basipterygoidal process Processus basitemporalis (Bock 1963)
Pi	Processus infracolumellaris extracolumellae (Gadow 1891)	Infracolumellar process Infrastapedial process (Smith 1905)

		Proc. infrastapedialis
		Proc. ventralis (NAA)
pm	Membrana postmeatica	
Po	Processus postorbitalis	
Poq	Processus oticus quadrati	Otic process of quadrate
Ps	Processus supracolumellaris (Gadow 1891)	Proc. suprastapedialis (Smith 1905)
		Proc. caudalis (NAA)
Psl	Processus supracolumellaris pars lateralis	Lateral prong of supracolumellar process
Psm	Processus supracolumellaris pars medialis	Medial prong of supracolumellar process
Pt	Processus tympanicus	Proc. extracolumellaris
		Proc. mediocolumellaris (Freye 1952/1953)
		Proc. extrastapedialis (Smith 1905)
		Proc. rostralis (NAA)
pto	Paratympanic organ	Organon vitali
Pz	Processus zygomaticus	Zygomatic process
Q	Quadratum	
Qca	Quadratocranial articulation	Articulatio quadrato-squamoso-otica (NAA2)
qJ	Quadrato-jugale	Quadratojugal bar
Ra	Radius	Barbule
ram	Ramus	Barb
Rc	Rachis pennae	Rachis
		Feather shaft
Rcst	Recessus scalae tympani	Perilymphatic recess
Rcta	Recessus tympanicus anterior	Anterior air space (Norberg 1978)
		Anterior tympanic recess
		Antrum pneumaticum rostrale (NAA)
		Posteroventral subcavity of the middle ear
		Presphenoid sinus (Saiff 1978)
		Recessus stapedialis (Freye-Zumpfe 1952/1953; Stresemann 1934)
		Recessus tympanicus anterior (Parker 1869; Suschkin 1899; Pycraft 1901; Müller 1963)
		Recessus tympanicus rostralis (NAA2)
		Sinus sphenoidalis (Pohlmann 1921)

Rctp	Recessus tympanicus posterior (Freye 1952/1953)	Posterior air space (Norberg 1978) Recessus tympanicus caudalis Recessus tympanicus inferior (Müller 1963)
Rcts	Recessus tympanicus superior (Freye 1952/1953)	Recessus tympanicus dorsalis (Suschkin 1899; Pycraft 1902; Müller 1963) Upper tympanic recess Superior air space (Norberg 1978)
RonV	Ramus ophthalmicus nervi trigemini	
Rps	Rostrum parasphenoidale	Rostroparasphenoid
rsnV_2	Ramus supraorbitalis nervi maxillaris trigemini	Supraorbital nerve of the maxillary branch of the trigeminal nerve
S	Stapes	Ossis columellae (Gadow 1891 cit. in Krause 1901)
Sc	Scala vestibuli	
Sca	Supracranial airspace	Cavum pneumaticum supracerebralis
sel	Subepidermal layer of eardrum	Connective tissue layer
Set	Setae	Bristles
Si	Siphonium	
Sp	Sinus pneumaticus marginalis (Pohlmann 1921; NAA)	
Sq	Squamosum	
stb	Stratum basale	
stc	Stratum corneum	Cornified layer of epidermis
Sti	Stipes (Krause 1901)	Scapus columellae Pars stapedialis ossis columellae (Krause 1901) Shaft of columella
stm	Stratum intermedium	Intermediate layer of epidermis
stmb	Secondary tympanic membrane	Membrana tympanica secundaria
Sy	Stylohyal	Stylohyal cartilage
tept	Tympanic epithelium of ear drum	
tm	Tympanic membrane	Membrana tympani (Freye 1952/1953) Membrana tympanica (NAA) Eardrum

Tpt	Tuba pharyngotympanica (NAA)	Eustachian tube Tuba Eustachii Tuba auditiva

NAA, *Nomina Anatomica Avium* (Baumel et al. 1979, 1993)

Springer-Verlag and the Environment

We at Springer-Verlag firmly believe that an international science publisher has a special obligation to the environment, and our corporate policies consistently reflect this conviction.

We also expect our business partners – paper mills, printers, packaging manufacturers, etc. – to commit themselves to using environmentally friendly materials and production processes.

The paper in this book is made from low- or no-chlorine pulp and is acid free, in conformance with international standards for paper permanency.